STUDY GUIDE

John A. Juve
Columbia College

An Introduction to

STATISTICS AND
RESEARCH METHODS

Becoming a Psychological Detective

STEPHEN F. DAVIS
Emporia State University

RANDOLPH A. SMITH
Kennesaw State University

PEARSON

Prentice
Hall

Upper Saddle River, New Jersey 07458

Printed in the United States of America
10 9 8 7 6 5 4 3 2 1

ISBN 0-13-150515-7

Table of Contents

Chapter 1

The Science of Psychology

Learning Objectives

By the end of this chapter you should be able to:

1. Summarize and use the methods for acquiring knowledge.

2. List and describe the components of the scientific method.

3. Recognize and describe the variables that comprise basic psychological experiments.

4. Discuss the factors related to establishing cause-and-effect relationships.

5. List and describe the steps involved with conducting a research study.

6. Describe the benefits of taking a statistics/research methods course.

To maximize your learning experience in this course, you need to become actively involved. One way to accomplish this task is to reflect on the timely and instructional **"Psychological Detective"** sections (hereafter symbolized as) in your book. Stop and reflect on these sections when you encounter them. Each section asks you to think about a question concerning psychological research. I recommend that you take full advantage of these sections because they are fun, challenging, and designed to help you think critically about psychological research. Consequently, I have highlighted each section for your easy reference in the expanded outlines.

Expanded Outline

I.   Ways to Acquire Knowledge
      Tenacity
      Authority

      Experience
      Reason and Logic
          Logical Syllogism
      Science

Practice Exam

Multiple Choice

Identify the letter of the choice that best completes the statement or answers the question.

____   1.   During World War II propaganda was continuously presented in the form of radio and newspaper. According to Charles S. Peirce, this method for acquiring knowledge is referred to as

        a. authority.   c. experience.
        b. tenacity.   d. reason and logic.

____   2.   Alex believes that four leaf clovers are good luck because he has repeatedly heard this from his brother. Alex has acquired his knowledge through

        a. experience.   c. reason and logic.
        b. tenacity.   d. none of the above

____   3.   According to the latest Tiger Woods TV commercial, the best golf shoes are made by Nike. This source of knowledge depends on

        a. tenacity.   c. credibility.
        b. authority.   d. knowledge of the product

____   4.   Heather's stockbroker has recommended that she purchase several high risk stocks. Although she is uncomfortable with this decision she agrees to purchase the stocks. Her decision is primarily influenced by

        a. tenacity.   c. reason and logic.
        b. experience.   d. authority.

3

_____ 5. Which of the following is <u>not</u> considered a problem associated with accepting an authority as a knowledge source?

a. there is no way to determine if the knowledge is accurate

b. the inability to change in the face of contradictory evidence

c. the credibility of the information source

d. the message must be repeated numerous times

_____ 6. If researchers asked 10 people to provide a description of a robbery they witnessed, they would likely obtain 10 different accounts of the event. Which of the following knowledge acquisition methods best describes this phenomenon?

a. tenacity

b. reason and logic

c. authority

d. experience

_____ 7. "Italian people love pizza. My friend Tony is from Italy, therefore, Tony must love pizza." This statement is an example of

a. tenacity.

b. inductive logic.

c. a logical syllogism.

d. a deductive syllogism.

_____ 8. Which of the following knowledge acquisition methods is preferred by researchers?

a. science

b. reason and logic

c. authority

d. experience

_____ 9. Which of the following is considered to mark the founding of psychology?

a. Titchener's use of introspection

b. Pavlov's work with classical conditioning

c. Wundt's research lab at the University of Leipzig

d. Freud's theory of personality

_____ 10. Researchers must use this component of the scientific method to ensure that the only factor that could influence their study is related to the independent variable (IV).

a. objectivity

b. self correction

c. confirmation of findings

d. control

4

____ 11. Researchers _____ the IV in a research study.

    a. manipulate   c. try to avoid the influence of
    b. measure or record   d. none of the above

____ 12. Researchers _____ the dependent variable (DV) in a research study.

    a. manipulate   c. try to avoid the influence of
    b. measure or record   d. none of the above

____ 13. Researchers _____ extraneous variables in a research study.

    a. manipulate   c. try to avoid the influence of
    b. measure or record   d. none of the above

____ 14. Annie conducts a study in which she examines the effects of previous coursework on college student achievement. In this study, the ethnicity of the participants could be considered a(n)

    a. DV.   c. extraneous variable.
    b. IV.   d. intrinsic variable.

____ 15. All good theories attempt to

    a. state specific IV and DV relations   d. both a and b
    b. point the way to new research   e. both b and c
    c. organize a given body of research

____ 16. Which of the following is not a key element in the scientific process?

    a. subjective measurements of the   c. self-correction of errors and
       phenomenon   faulty reasoning
    b. ability to verify or confirm previous d. exercising control to rule out
       measurements   the influence of unwanted
                                 factors

____ 17. Professor Parker administers an attitude scale to participants who have been sitting for one hour in either a dark or well lit classroom. What is the DV in this experiment?

    a. the professor   c. time in classroom
    b. the attitude scale   d. dark or well lit classroom

_____ 18. Psychologists use the term _____ to refer to a research study that is conducted in exactly the same manner as a previous study.

    a. subjectivity   c. empirical
    b. objectivity   d. replication

_____ 19. Researchers are able to infer a cause-and-effect relationship when they have

    a. controlled potential extraneous variables.   c. both a and b
    b. manipulated an IV.   d. none of the above

_____ 20. The use of specific IV and DV relations subsumed under a larger research area best describes

    a. replication.   c. a theory.
    b. replication with extension.   d. a hypothesis.

_____ 21. A(n)_____is referred to as the predicted outcome of the research project.

    a. experimental hypothesis   c. logical syllogism
    b. theory   d. correlation

_____ 22. Which of the following components of a research process typically occurs at the end __and__ at the beginning of a research process representing a full research cycle?

    a. literature review   c. analyzing the data
    b. finding a new problem   d. presenting your results

_____ 23. Art's statistical test has indicated that there was a significant difference between his two research groups. He is currently in the process of

    a. sharing the results.   c. conducting the experiment.
    b. analyzing his data.   d. preparing the research report.

_____ 24. Which of the following is a method for sharing or presenting the results of your research?

    a. oral papers   c. professional research journals
    b. posters   d. all of the above

Matching

a. continued presentation of a
   particular bit of information

b. a method of knowledge acquisition
   preferred by researchers

c. objectively quantifiable measurements

d. a study conducted the same
   way as a previous study

e. contains a faulty assumption

f. a manipulated variable

g. a measured variable

h. an unwanted variable

i. formal statement of the relations
   among the IVs and DVs

j. specific statement about IV
   and DV relations

k. predicted outcome of a research study

l. general plan for conducting research

_a_ 1. tenacity

_d_ 2. replication

_f_ 3. IV

_j_ 4. hypothesis

_b_ 5. science

_l_ 6. research design

_h_ 7. extraneous variable

_c_ 8. empirical

_k_ 9. experimental hypothesis

_e_ 10. logical syllogism

_g_ 11. DV

_i_ 12. theory

True/False

Indicate whether the sentence or statement is true or false.

F   1.   The continued presentation of a particular bit of information is known as a logical syllogism.

T   2.   Empirical measurements are based on objectively quantifiable observations.

T   3.   Although knowledge may be obtained through several sources, the science method is preferred.

F   4.   Researchers seek to establish cause-and-effect relations between IVs and DVs by controlling extraneous variables.

F   5.   Logical syllogisms are founded on accurate assumptions.

T   6.   Theories attempt to state specific IV and DV relations.

T   7.   Researchers use the American Psychological Association's Publication Manual as a guideline for preparing written research reports.

F   8.   Independent variables are responses or behaviors that researchers measure.

T   9.   The experimenter's predicted outcome of a research project is referred to as the research or experimental hypothesis.

F   10.   Undergraduate students are <u>not</u> allowed to publish their research in professional journals.

Short Answer

1. Describe the problems associated with the tenacity and authority sources of knowledge. How do these problems influence the ways that researchers acquire knowledge?

2. Briefly define logical syllogism. Summarize the problems associated with this type of knowledge source. Create your own logical syllogism.

3. List and discuss the purpose of the four elements of the scientific process.

4. Explain why it is important to replicate previous research findings. Summarize why the method of replication with extension can benefit researchers?

5. Describe the roles that IVs, DVs, and extraneous variables play in designing and conducting research studies. Create a short study and identify each of the above variables. Briefly discuss their roles in your study.

6. Explain the relevance of the phrase "self-correcting nature of science" to the research process.

7. Discuss two ways in which researchers can assert control in a research project.

8. List and describe each step in the research process. Draw a conceptual map or diagram by using figures and arrows to display the sequential steps of the research process.

9. Provide several examples of forums that allow students the opportunity to present and publish the results of their research.

10. You have been asked by your professor to recruit students for his research methods/statistics course. Prepare a short list of reasons discussing why it is beneficial for students to take his course. Try to emphasize how this course will help the students in the future.

Answers to the multiple choice, matching, and true/false items

Multiple Choice		Matching		True and False	
1.	B				
2.	B	1.	A	1.	F
3.	C	2.	D	2.	T
4.	D	3.	F	3.	T
5.	C	4.	J	4.	T
6.	D	5.	B	5.	F
7.	C	6.	L	6.	F
8.	A	7.	H	7.	T
9.	C	8.	C	8.	F
10.	D	9.	K	9.	T
11.	A	10.	E	10.	F
12.	B	11.	G		
13.	C	12.	I		
14.	C				
15.	E				
16.	A				
17.	B				
18.	D				
19.	C				
20.	D				
21.	A				
22.	B				
23.	B				
24.	D				

Chapter 2

Research Ideas and Hypotheses

Learning Objectives

By the end of this chapter you should be able to:

1. Summarize the nature of research ideas and hypotheses.

2. Describe the characteristics of and sources for good research ideas.

3. Develop a research question.

4. Conduct a computerized review of the research literature.

5. Discuss strategies for obtaining relevant publications.

6. Integrate the results of the literature review.

Expanded Outline

I. The Research Idea
   Characteristics of Good Research Ideas
      Testable
      Likelihood of Success

   Sources of Research Ideas
      Nonsystematic
         Inspiration
         Serendipity
         Everyday Occurrences
      Systematic
         Past Research
         Theory
         Classroom Lectures

II. Developing a Research Question

Practice Exam

Multiple Choice

Identify the letter of the choice that best completes the statement or answers the question.

____ 1. The most important characteristic of a good research idea is

   a. inspiration.   c. serendipity.
   b. testability.   d. likelihood of success.

____ 2. The likelihood of success in a research study is best associated with

   a. reality.   c. a hypothesis.
   b. testability.   d. a theory.

____ 3. The two primary sources of research ideas are

   a. nonsystematic and inspirational. c. systematic and testability.
   b. nonsystematic and testability.   d. systematic and nonsystematic.

____ 4. According to Koestler (1964), one of the most famous examples of a nonsystematic research idea involved ideas popping into Albert Einstein's brain while he was sailing. Which nonsystematic research source does this example illustrate?

   a. altruism   c. serendipity
   b. inspiration   d. everyday occurrences

____ 5. All of the following are examples of systematic sources of research ideas except

   a. classroom lectures.   c. past research.
   b. theories.   d. inspiration.

____ 6. Classroom lectures are considered to be systematic sources of research ideas because they

   a. are presented by someone   c. include an organized review
      familiar with the content.   of relevant research.
   b. occur on a regular basis   d. often include
      throughout the semester.   laboratory experiences.

_____ 7. If you are searching for terms related to the word "mania" you will most likely use

    a. the thesaurus of psychological terms.   c. Psychological Abstracts.
    b. a dictionary of psychology terms.   d. PsychLIT.

_____ 8. Researchers will most likely use PsycINFO when they are

    a. obtaining relevant publications.   c. manually searching the literature.
    b. conducting computer searches   d. integrating the results of the
       of the literature.   literature search.

_____ 9. According to your book, it is important to make your literature review as thorough as possible to

    a. avoid repeating previous mistakes.   d. both a and c
    b. facilitate the manuscript page length requirement.   e. both b and c
    c. avoid replicating previous research.

_____ 10. The terms "authority," "accuracy," "objectivity," "currency," and "coverage" are often used to evaluate

    a. Internet resources.   c. peer reviewed journals.
    b. psychological abstracts.   d. library books.

_____ 11. Annie has created a list of books and journal articles that she needs to begin her research project. She learns that several of the sources on her list are not available at the local library. Which of the following options would you recommend for her?

    a. search for an alternative source   c. author reprints
    b. interlibrary loan   d. both b and c

_____ 12. While reading about a psychology experiment you find yourself paying attention to a description of the participants used in the experiment. Which section of the research paper are you reading?

    a. results   c. method
    b. discussion   d. abstract

_____ 13. In which section of a research paper would you most likely find a description of the materials or equipment used by a researcher?

    a. procedure   c. discussion
    b. results   d. apparatus

_____ 14. You predict that rats raised in an enriched or stimulating environment will learn to complete complex tasks faster than rats raised in a non-enriched or isolated environment. Which step in the research process are you conducting?

    a. surveying the literature   c. conducting the experiment
    b. developing a good research idea d. formulating the hypothesis

_____ 15. The statement, "I am a great golfer or I am not a great golfer," is an example of a(n) _____.

    a. analytic statement   c. contradictory statement
    b. synthetic statement   d. inductive statement

_____ 16. "I am passing my statistics course and I am not passing my statistics course" is an example of the following type of statement

    a. analytic.   c. contradictory.
    b. synthetic.   d. inductive.

_____ 17. A good research hypothesis can be either true or false, and is therefore a(n)_____ type of statement.

    a. analytic   c. contradictory
    b. synthetic   d. deductive

_____ 18. Which of the following statements should researchers use to properly construct a sentence with general implication form?

    a. "or and then"   c. "or and because"
    b. "if and then"   d. "if and because"

_____ 19. If children brush their teeth after every meal, then they will have fewer cavities than children who do not brush their teeth after every meal. What type of statement does this example illustrate?

    a. analytic   c. rhetorical
    b. contradictory   d. general implication form

_____ 20. When using the general implication format, the "if" portion refers to the manipulation of the _____; whereas the "then" portion refers to changes in the _____ variable that we expect to observe.

    a. dependent, extraneous   c. independent, nuisance
    b. independent, dependent   d. dependent, independent

_____ 21. Researchers must be able to state research hypotheses in this type of format to conduct a proper study.

    a. analytical
    b. extraneous
    c. general implication
    d. contradictory

_____ 22. What principle are we utilizing if our experiment does not turn out as predicted and our result is viewed as evidence that our hypothesis is false?

    a. truth
    b. attribution error
    c. falsifiability
    d. alternative hypothesis

_____ 23. Reasoning from general principles to specific conclusions or predictions best describes

    a. deductive logic.
    b. intuitive logic.
    c. inductive logic.
    d. the principle of falsifiability.

_____ 24. Reasoning from specific cases to general principles best describes

    a. deductive logic.
    b. intuitive logic.
    c. inductive logic.
    d. the principle of falsifiability.

_____ 25. Researchers utilize _____ logic while formatting research hypotheses and _____ logic while constructing theories.

    a. inductive, deductive
    b. inductive, reductive
    c. reductive, deductive
    d. deductive, inductive

_____ 26. Predicting that students who have taken extensive math and science courses will score significantly higher on the Medical College Admissions Test than students who have taken only the required math and science courses is an example of a(n) _____ hypothesis.

    a. directional
    b. non-directional
    c. reductive
    d. inductive

_____ 27. Thurstone predicts that female students will perform differently than male students on his English exam. His hypothesis is an example of a(n)

    a. directional hypothesis.
    b. nondirectional hypothesis.
    c. reductive hypothesis.
    d. inductive hypothesis.

Matching

a. synthetic   h. interlibrary loan

b. method section   i. research idea

c. analytic   j. inductive logic

d. inspiration   k. serendipity

e. testable   l. classroom lecture

f. deductive logic   m. PsycLIT

g. contradictory   n. results

___c___ 1. statements that can be either true or false

___j___ 2. process used to formulate a research hypothesis

___i___ 3. identification of a gap in knowledge

___k___ 4. searching for phenomenon, but finding something else

_____ 5. statements that are always true

_____ 6. statements that are always false

___l___ 7. systematic research source

___m___ 8. used to conduct a computerized search of the literature

___d___ 9. nonsystematic research source

___n___ 10. section of a research paper where statistical tests are discussed

___b___ 11. section of a research paper where participants are discussed

___h___ 12. method for obtaining relevant publications

___f___ 13. process used to construct theories

___e___ 14. characteristic of a good research idea

17

True/False

Indicate whether the sentence or statement is true or false.

_____ 1. The most important characteristic of a good research idea is that it is consistent.

_____ 2. Serendipity refers to situations where we look for one phenomenon but find another.

_____ 3. PsycFIND helps researchers to conduct computerized searches of the literature.

_____ 4. Researchers can locate descriptions about the equipment used in a research project in the results section.

_____ 5. Synthetic statements are always true.

_____ 6. Deductive logic is a reasoning process used when formulating a research hypothesis.

_____ 7. The statement, "the responses of men will significantly differ from the responses of women," is an example of a directional hypothesis.

_____ 8. The predicted outcome of a research project is referred to as the Principle of Falsifiability.

_____ 9. Analytic statements are always true.

_____ 10. Inspiration is considered a major systematic source for research ideas.

Short Answer

1. List and summarize the characteristics of good research ideas.

2. Discuss the two general sources for obtaining research ideas?

3. Explain how understanding past research can improve future research.

4. Your study has failed to replicate a previous experiment. Does this mean your study was not meaningful? Explain your answer.

5. List and describe the steps involved with conducting a literature review.

6. Discuss the primary difference between PsychINFO and PsycFIRST.

7. Provide several reasons why researchers should be critical of Internet resources.

8. List and describe the common components of a research article.

9. Discuss the role that the Principle of Falsifiability plays in research.

10. Distinguish between inductive and deductive logic. Can they be used simultaneously by researchers? Explain your response.

Answers to the multiple choice, matching, and true/false items

Multiple Choice		Matching		True and False	
1.	B	1.	A	1.	F
2.	A	2.	F	2.	T
3.	D	3.	I	3.	F
4.	B	4.	K	4.	F
5.	D	5.	C	5.	F
6.	C	6.	G	6.	T
7.	A	7.	L	7.	F
8.	B	8.	M	8.	F
9.	D	9.	D	9.	T
10.	A	10.	N	10.	F
11.	D	11.	B		
12.	C	12.	H		
13.	D	13.	J		
14.	D	14.	E		
15.	A				
16.	C				
17.	B				
18.	B				
19.	D				
20.	B				
21.	C				
22.	C				
23.	A				
24.	C				
25.	D				
26.	A				
27.	B				

Chapter 3

Ethics in Psychological Research

Learning Objectives

By the end of this chapter you should be able to:

1. Discuss the rationale for establishing ethical principles in psychological research.

2. Summarize the APA ethical principles for conducting research with humans and animals.

3. Describe the purpose and functions of the Internal Review Board (IRB).

4. Explain the responsibilities that participants and experimenters have while conducting research.

5. Discuss the ethical obligations that researchers have after an experiment has been completed.

Expanded Outline

    I.   The Need for Ethical Principles

    II.   APA Principles in the Conduct of Research with Humans

            Is Deception in Research Necessary?
            Informed Consent

            Participants at Risk and Participants at Minimal Risk
                  Participants at Risk
                  Participants at Minimal Risk
             Vulnerable Populations
            The Debriefing Session

    III.   Ethical Use of Animals in Psychological Research

    IV.   The Institutional Review Board (IRB)

    V.   The Experimenter's Responsibility

VI.   The Participant's Responsibility

VII.   Researchers' Ethical Obligations Once the Research is Completed
            Plagiarism
            Fabrication of Data

            Lying with Statistics
            Citing Your References Correctly

VIII.   Review Summary

IX.   Check Your Progress

X.   Key Terms

XI.   Looking Ahead

Practice Exam

Multiple Choice

Identify the letter of the choice that best completes the statement or answers the question.

_____ 1.   Under the 1974 National Health Research Act, the US government demands
             assurance that federally funded projects have

             a. adequate funding for safe materials.   c. obtained informed consent.
             b. been reviewed and approved by peers.   d. both b and c

_____ 2.   During WWII Nazi doctors used unethical research practices to conduct
             experiments on their prisoners. These practices eventually led to the development
             of a code of medical and research ethics called the

             a. American Psychology Guidelines.   c. American Science Code.
             b. Nuremburg Code.   d. American Educational Guidelines.

_____ 3. The incidents at the Willowbrook Mental Health Facility were considered unethical because the

    a. patients' parents did not consent.   c. patients' were infected and not treated.
    b. drugs were not tested.   d. staff was not qualified to conduct research.

_____ 4. Which ethical research aspect was <u>not</u> stressed by the Nuremberg Code?

    a. research conducted by qualified staff   c. the right to discontinue participation
    b. must be 18 years old to participate   d. the right to be fully informed

_____ 5. In the Tuskegee Syphilis Study, _____ were recruited as participants.

    a. mentally ill patients   c. African-American men
    b. children   d. African-American women

_____ 6. Which of the following principles of the Nuremberg Code was <u>not</u> violated in the Tuskegee Syphilis Study?

    a. research conducted by qualified staff   c. the right to be fully informed
    b. the right to consent to participate   d. the right to discontinue participation

_____ 7. In 1963 Stanley Milgram conducted an obedience-to-authority study. However, contrary to his initial prediction, the participants in the study administered _____ shocks at what they assumed were _____ voltage levels.

    a. a few, very low   c. a few, very high
    b. several, very low   d. several, very high

_____ 8. An informed consent form should address all of the following issues <u>except</u>

    a. the general description of the study.   c. the right to withdraw without penalty.
    b. the right to discontinue treatment.   d. the statistical tests used to analyze data.

_____ 9. The primary purpose of a debriefing session is to

    a. explain the study and alleviate   c. identify participants at risk.
       undesirable consequences.
    b. recite the informed consent form.   d. discuss the conclusions of the study.

_____ 10. In what year did Willard Small use the first rat maze at Clark University?

    a. 1850   c. 1900
    b. 1875   d. 1925

____ 11. The primary role of the Institutional Review Board (IRB) involves

     a. determining the merits of a study.   c. editing manuscripts before publication.
     b. ensuring participants are treated   d. conducting debriefing sessions.
        ethically.

____ 12. Alex wants to conduct a research project with white mice. The IRB committee that will review his research proposal will likely consist of

     a. faculty members.   c. a veterinarian.
     b. individuals from the community.   d. all of the above

____ 13. Ultimately, who is accountable for the ethical conduct of a student research project?

     a. the student's advisor   c. the IRB
     b. the student   d. the psychology department

____ 14. According to the Department of Psychology at Bishop University (1994), which of the following suggestions is <u>not</u> accurate when trying to avoid plagiarism?

     a. never submit previous work for   c. reference statements that are not yours
        another class   unless it is common knowledge
     b. it is not necessary to acknowledge d. when using the exact words of another
        secondary sources   author you must use quotation marks

____ 15. Possibly the most famous instance of apparent data fabrication involved

     a. Cyril Burt.   c. Darrell Huff.
     b. Sherlock Holmes.   d. Willard Small.

____ 16. Fabrication of data refers to those instances where the experimenter deliberately

     a. submits previous work for another class. d. a and b
     b. changes data.   e. b and c
     c. makes up data.

____ 17. In 1954 Darrell Huff published a book describing questionable statistical practices designed to influence the conclusions reached by the readers. What was the name of his book?

     a. *Presenting Biased Results*   c. *How To Lie With Statistics*
     b. *Identifying Questionable Statistics* d. *The Art Of Data Cleaning*

_____ 18.   Heather has read about an interesting and relevant piece of research and she is having difficulty acquiring a copy of the original report. Her library does not carry the journal in which the article appears and she cannot obtain the article using interlibrary loan. Lastly, she cannot find the author's address. What should she do in this situation?

  a. cite only the original article c. search for another research idea
  b. cite the secondary source   d. none of the above

_____ 19.   Which of the following organizations could be used when requesting permission to conduct an experiment?

  a. Animal Care and Utilization Committee c. Institutional Review Board
  b. Human Subjects Review Panel   d. all of the above

_____ 20.   Sadie has just completed her research about the influence of Greyhound dogs on elderly depression. What ethical responsibilities should she be aware of now that her research is completed?

  a. obtaining informed consent d. both a and b
  b. plagiarism   e. both b and c
  c. fabrication of data

Matching

a. Willard Small   f. teachers and learners

b. ethical guidelines   g. plagiarism

c. mentally ill patients   h. Darrell Huff

d. children   i. debriefing session

e. IRB   j. African-American men

_a_ 1. used the first rat maze in 1900

_b_ 2. Nuremberg Code

_j_ 3. participants in the Tuskegee Syphilis study

_e_ 4. review panel for the use of human participants

_g_ 5. failure to acknowledge secondary sources

_i_ 6. final step in conducting the research project

_c_ 7. Willowbrook participants

_h_ 8. wrote a book entitled *How To Lie With Statistics*

_f_ 9. participants in Milgram's obedience-to-authority study

_d_ 10. is considered to be a vulnerable population

True/False

Indicate whether the sentence or statement is true or false.

_____ 1. The experimenter's ethical responsibilities end when the data are collected and the participants are debriefed.

_____ 2. The Institutional Review Board (IRB) <u>only</u> reviews research proposals pertaining to human participants.

_____ 3. The main purpose of the IRB is to explain the nature of an experiment and remove any undesirable consequences.

_____ 4. Although researchers should try to avoid it, using deception may be justified in order to yield unbiased responses.

_____ 5. A debriefing session typically occurs before a research experiment.

_____ 6.   When a study has been completed, participants share the responsibility for understanding what happened.

_____ 7.   Failing to acknowledge secondary sources is <u>not</u> a violation of plagiarism.

_____ 8.   It is the researcher's ethical responsibility to cite and list only those works that have been read.

_____ 9.   It is possible to lie with statistics.

_____ 10.   The Luxemburg Code explains that research projects should be conducted by scientifically qualified personnel.

Short Answer

1.   Summarize the ethical research considerations stressed by the Nuremberg Code. Explain the events leading up to the development of this code.

2.   Explain why the APA ethical research standards pertaining to securing informed consent and using deception in research are controversial? Is deception in research necessary or justified? Provide a rationale for your answer.

3.   Distinguish between "participants at risk" and "participants at minimal risk."

4.   Discuss how debriefing sessions can benefit both researchers and participants?

5.   Do participants have ethical responsibilities? What should researchers expect from their participants?

6.   As part of the debriefing session, the researcher may ask the participants to not discuss the experiment with their peers. Discuss why this is important. How can non-compliance with this request influence the research project?

7.   According to the Department of Psychology at Bishop University (1994), what can you do to avoid plagiarism? Try to list and describe several suggestions for avoiding plagiarism.

8.   Summarize why is it important to acknowledge secondary sources in your literature review. Discuss why this is related to plagiarism.

9.   Explain why a researcher would engage in plagiarism or fabrication of data.

10.   Describe what is meant by the statement "lying with statistics."

Answers to the multiple choice, matching, and true/false items

Multiple Choice	Matching	True and False
1. D	1. A	1. F
2. B	2. B	2. F
3. C	3. J	3. F
4. B	4. E	4. T
5. C	5. G	5. F
6. A	6. I	6. T
7. D	7. C	7. F
8. D	8. H	8. T
9. A	9. F	9. T
10. C	10. D	10. F
11. B		
12. D		
13. B		
14. B		
15. A		
16. E		
17. C		
18. B		
19. D		
20. E		

Chapter 4

Basic Statistical Concepts, Frequency Tables, Graphs, Frequency Distributions, and Measures of Central Tendency

Learning Objectives

By the end of this chapter you should be able to:

1. Define and describe descriptive and inferential statistical procedures.

2. Discuss when and why descriptive and inferential procedures are used.

3. Distinguish between nominal, ordinal, interval, and ratio scales.

4. Create and interpret ungrouped and grouped frequency distributions when reporting and discussing statistical findings.

5. Summarize when it is appropriate to use pie charts, bar graphs, histograms, frequency polygons, and line graphs to explain your data.

6. Describe the shapes of frequency distributions.

7. List and define the measures of central tendency.

8. Calculate and interpret the measures of central tendency.

9. Compare and contrast the usefulness of central tendency measures in applied situations.

10. Identify and describe the shape of a distribution with respect to its skewness.

Expanded Outline

    I.   Two Branches of Statistics
                Descriptive
                Inferential

    II.   Measurement
                Scales of Measurement
                    Nominal
                    Ordinal
                    Interval

                    Ratio

III.   Frequency Distributions

        Grouped Frequency Distributions

IV.   Review Summary

V.   Check Your Progress

VI.   Graphing Your Results
        Pie Chart
        Bar Graph
        Histogram

        Frequency Polygon
        Line Graph
            Ordinate
            Abscissa

VII.   Frequency Distributions and Their Shapes
        Unimodal
        Bimodal
        Multimodal

        Rectangular Distribution

VIII.   Measures of Central Tendency
        Mode
        Median
        Mean
        Choosing a Measure of Central Tendency
            Skewed Distributions
                Positive
                Negative

IX.   Review Summary

X.   Check Your Progress

XI.   Exercises

XII.   Key Terms

XIII.   Looking Ahead

Practice Exam

Multiple Choice

Identify the letter of the choice that best completes the statement or answers the question.

_____ 1. Statistics is a branch of mathematics that involves

    a. collecting data.
    b. analyzing data.
    c. interpreting data.
    d. all of the above

_____ 2. What are the basic divisions of statistics?

    a. sampling and scaling
    b. inferential and descriptive
    c. population and sample
    d. statistics and parameter

_____ 3. _____ statistics are methods used to clarify and summarize numerical data, whereas_____ statistics are methods used to make generalizations about a population by studying a sample of that population.

    a. Descriptive, interval
    b. Descriptive, inferential
    c. Inferential, ratio
    d. Inferential, descriptive

_____ 4. A researcher hypothesizes that children who have different eye colors also have different IQ scores. Eye color is an example of what type of measurement scale?

    a. nominal
    b. ordinal
    c. interval
    d. ratio

_____ 5. A university scholarship foundation obtains the following information on students applying for financial assistance: sex, age, academic department, SAT scores, and cumulative grade point average. What measurement scale would be used to describe the variable of sex?

    a. nominal
    b. ordinal
    c. interval
    d. ratio

_____ 6. Given the information in question #5, what measurement scale would be used to describe the age variable?

    a. nominal
    b. ordinal
    c. interval
    d. ratio

_____ 7. Given the information in question #5, what measurement scale would be used to describe the SAT variable?

    a. nominal   c. interval
    b. ordinal   d. ratio

_____ 8. A tabulation of data that indicates the number of times a given score or groups of scores appear is known as a

    a. sample.   c. rank distribution.
    b. frequency distribution.   d. parameter.

_____ 9. Which type of graph presents data in terms of frequency per category and is utilized when researchers use nominal or qualitative categories

    a. bar graph.   c. histogram.
    b. pie chart.   d. frequency polygon.

_____ 10. When constructing a line graph, we start with two axes or dimensions. The vertical or Y axis is know as the _____ and the horizontal or X axis is known as the _____ .

    a. abscissa, slope   c. ordinate, abscissa
    b. abscissa, ordinate   d. ordinate, intercept

_____ 11. Which of the following is not a measure of central tendency?

    a. mode   c. mean
    b. standard deviation   d. median

_____ 12. What is the term for the number or score that divides the distribution into equal halves and is referred to as the 50th percentile?

    a. mode   c. mean
    b. standard deviation   d. median

_____ 13. Consider the following scores for questions 13, 14, and 15:

[18, 20, 16, 12, 20, 13, 24, 17, and 19].

What is the mode for this distribution of scores?

    a. 16   c. 17
    b. 20   d. 19

_____ 14. What is the median for this distribution of scores?

    a. 18   c. 19
    b. 17   d. 16

_____ 15. What is the mean for this distribution of scores?

    a. 17.00   c. 18.20
    b. 16.66   d. 17.66

_____ 16. We decide which measure of central tendency to use based on the _____ used to measure the variable and the _____ of the distribution.

    a. shape, ratio   c. scale, shape
    b. shape, scale   d. scale, ratio

_____ 17. A nonsymmetrical distribution is known as a

    a. skewed distribution.   c. frequency distribution.
    b. normal distribution.   d. none of the above

_____ 18. A skewed curve with its tail to the right is described as

    a. negatively skewed.   c. symmetrically skewed.
    b. positively skewed.   d. normally skewed.

_____ 19. In a positively skewed distribution, the mean is _____ than the median, and the median is _____ than the mode.

    a. smaller, smaller   c. larger, smaller
    b. smaller, larger   d. larger, larger

_____ 20. In a negatively skewed distribution, the _____ has the smallest value, the _____ has the largest value, and the _____ falls between them.

    a. mean, median, mode   c. median, mode, mean
    b. mean, mode, median   d. mode, mean, median

Matching

a. abscissa   f. skewed

b. interval   g. ordinal

c. nominal   h. graph

d. bar chart   i. histogram

e. ratio   j. ordinate

____ 1. loudness

____ 2. a nonsymmetrical distribution

____ 3. presents qualitative data in terms of frequencies per category

____ 4. pictorial representation of a set of data

____ 5. the vertical or Y axis of a graph

____ 6. presents quantitative data in terms of frequencies per category

____ 7. the horizontal or X axis of a graph

____ 8. categories of aggressiveness

____ 9. political party membership

____ 10. Fahrenheit temperature

True/False

Indicate whether the sentence or statement is true or false.

____ 1. A graph is an example of a descriptive statistic.

____ 2. The answers we obtain from descriptive procedures are called statistics.

____ 3. When we use numbers to identify a quality or category, we have an ordinal scale.

____ 4. When numbers measure an amount, but there is no true zero, we have an interval scale.

____ 5. Researchers often create a grouped frequency distribution to demonstrate how often each score occurred.

____ 6. The DV is plotted on the ordinate.

____ 7. The IV is plotted on the ordinate.

_____ 8.   Strictly speaking a binomial distribution may also be referred to as a multimodal distribution.

_____ 9.   A central value, between the extreme scores in a distribution, around which the scores are distributed, is called variation.

_____ 10.   A negatively skewed curve has its tail to the left.

Short Answer

1.   Explain why researchers use statistics.

2.   Distinguish between nominal, ordinal, interval, and ratio scales.

3.   Compare and contrast descriptive and inferential statistics.

4.   List and explain the steps used to construct a frequency distribution.

5.   Distinguish between unimodal, bimodal, and multimodal distributions?

6.   Summarize the information that is needed to determine which measure of central tendency you should use to describe your data.

7.   Explain what a measure of central tendency tells us about a distribution.

8.   What aspects of data determine which measures of central tendency we will use to describe our data?

9.   Discuss why the mean is considered to be an inappropriate measure of central tendency in an extremely skewed distribution.

10.   If the median of a distribution is considerable greater than the mean, what do you know about the shape of the distribution?

SPSS Computer Practice Problem 4.1 Using SPSS 11.0 for Windows

Basically when conducting a statistical analysis using SPSS for Windows you will name your variables, enter your data, and finally analyze your data by selecting options from a toolbar. In the following section a step by step procedure of an example problem will be provided to guide you in creating a frequency table and histogram. By following the systematic instructions provided and by referring to the screen figures (SF) when prompted, you should be able to independently create a frequency table and histogram.

Example Computer Problem 4.1

Data
The data for this problem is based on a 20-point statistics quiz administered to 20 students on the first day of class. The raw scores for each of the students are provided below:

4, 20, 0, 7, 7, 3, 10, 4, 6, 20, 5, 9, 6, 8, 7, 5, 8, 5, 6, and 8

Assignment
1. Create a <u>frequency table</u> and a <u>histogram</u> based on the raw data for the student quiz scores.
2. Request that SPSS report the following descriptive statistics:
   - mean, mode, median
   - range
   - standard deviation
   - skewness
3. Describe the shape of the histogram distribution. Are there outliers in this data set?
4. Explain what measure of central tendency best describes this data?

Procedures for Completing Example Problem 4.1

1. **Getting Started.** Open the SPSS 11.0 program
2. **Data Option.** At this point the SPSS window should provide a prompt asking, "What would you like to do?" Using your mouse click on the radial button to "Type in data."
3. **Data Entry.** Next you will begin to enter the data. This requires manually typing each raw score into a column (see Screen Figure 4.1 or SF-4.1). For example, you will enter the first student's quiz score (4) in the first data cell for column one under the variable category "var00001." Next you will enter the second student's quiz score (20) in the second data cell for column one and so on until you have entered all 20 quiz scores.

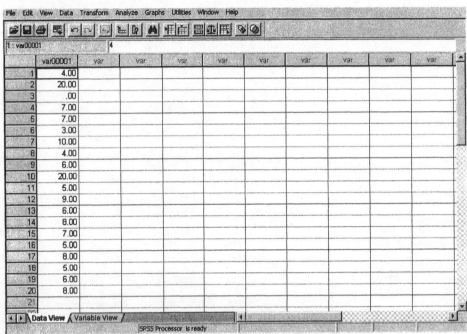

SF-4.1. Entering data to conduct a frequency distribution.

4. **Naming Your Variable(s).** At this point it may be helpful to rename or replace the default variable "var00001" by selecting or clicking on the variable view tab at the bottom left hand corner of the screen (see SF-4.1). Naming your variables will be especially important for organization and record keeping when you begin to add more variables to your data set. After selecting the variable view tab you should be able to view the variable named "var00001" (see SF-4.2). Under the column "Name" in the first cell you can provide an alternative name for your variable by highlighting or selecting the cell with your mouse. At this point your variable cell should be highlighted with a black box (see SF-4.2). You may use up to eight characters to name your variable. Each horizontal row on the left hand side of the screen, indicated by sequential numbering, represents a potential variable. For the purpose of this example the variable will be named "Quiz" to more accurately explain the data. Once finished, select the "Data View" tab at the lower left hand corner of the screen and this will return you to the screen displaying your raw data.

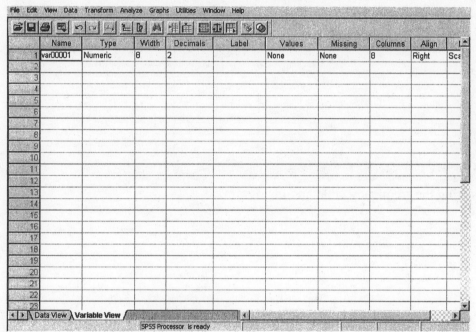

SF-4.2. Naming your variable.

5. **Creating the Frequency Table and Histogram.** In order to conduct a statistical analysis using SPSS you will need to select the "Analyze" toolbar option at the top of the page. A dropdown box will appear providing further data analysis options. Select the "Descriptive Statistics" option. After selecting this option SPSS will provide you with further analysis alternatives to the right of the initial dropdown box (see SF-4.3).

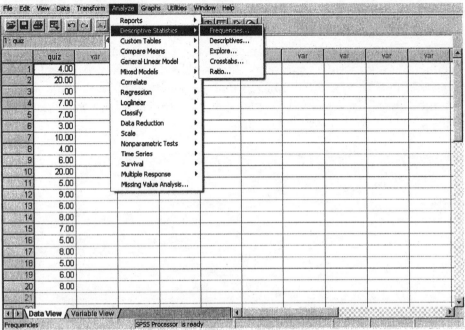

SF-4.3. Creating the frequency table and histogram.

Next select the "Frequencies" option. At this point you should be able to view a smaller command screen labeled "Frequencies" (see SF-4.4). Select or highlight the variable (in this case quiz) by clicking on the variable name and then use the arrow button located in the middle of the screen to move your variable into the variable box.

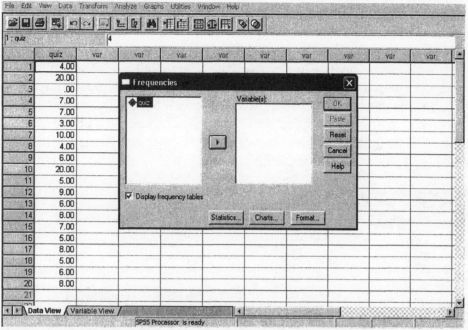

SF-4.4. Creating the frequency table and histogram: Selecting a variable.

After selecting your variable you will need to make some choices concerning what type of descriptive and graphical output you will need to appropriately describe your data. First you <u>must</u> select the "Display frequency tables" option in the lower left hand corner of the original "Frequencies" command screen. If this option is not selected SPSS will not provide the output you need. Next select the "statistics" button located in the middle bottom portion of the "Frequencies" command screen. Once this option is selected a second command screen will appear titled "Frequencies: Statistics" which can be viewed in the lower right hand corner of Screen Figure 4.5. This command screen will provide you with several options to describe your data including: "Percentile Values," "Dispersion," "Central Tendency," and "Distribution." For the purpose of this example select the following options as shown in Screen Figure 4.5: "Std (standard) deviation," "Range," "Mean," "Median," "Mode," and "Skewness." Then click on the "Continue" button located in the upper right hand corner of the "Frequency: Statistics" command screen which will prompt the screen to disappear.

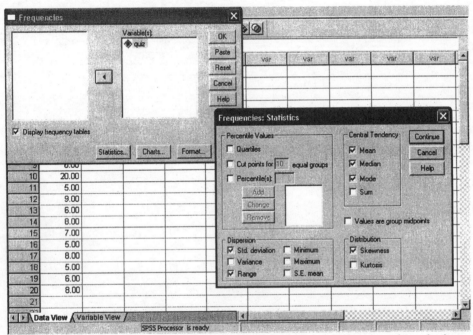

SF-4.5. Creating the frequency table and histogram: Selecting descriptive statistics

Next you will need to choose a chart to graphically display your statistical output and this can be accomplished by selecting the "Charts" option located in the middle of the "Frequencies" command screen. After selecting this option another command screen titled "Frequencies: Charts" will appear as shown in the lower right hand corner of Screen Figure 4.6. This screen allows you to select from Bar and Pie charts as well as Histograms. For the purpose of this example select the "Histograms" option (see SF-4.6). Then click on the "Continue" button located in the upper right hand corner of the "Frequency: Charts" command screen which will prompt the screen to disappear.

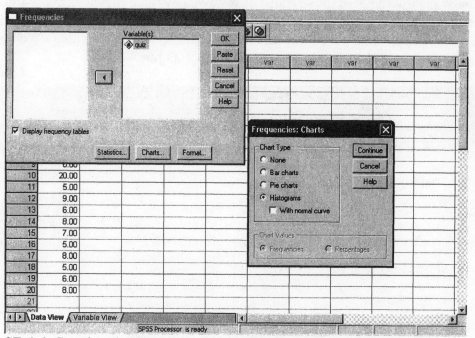

SF-4.6. Creating the frequency table and histogram: Selecting a chart.

At this point you will once again see the "Frequencies" command screen, as shown in Screen Figure 4.4, however your selected variable should be located in the "Variable(s)" box. Next select the "OK" command button located in the upper right hand corner of the screen. This will prompt SPSS to begin running your analysis.

6. **Reviewing the Descriptive and Graphical Output**. After the frequency analysis has been completed by SPSS an output screen titled "Output1" will appear (see SF-4.7). The title will be displayed in the top left corner of screen figure 4.7. This screen will sequentially present the descriptive statistics followed by the frequency counts and the histogram as indicated in the upper left hand output column of Screen Figure 4.7. Given that all of the output could not be displayed in Screen Figure 4.7, the complete output for this analysis will be provided in Tables 4.1, 4.2 and in Figure 4.1.

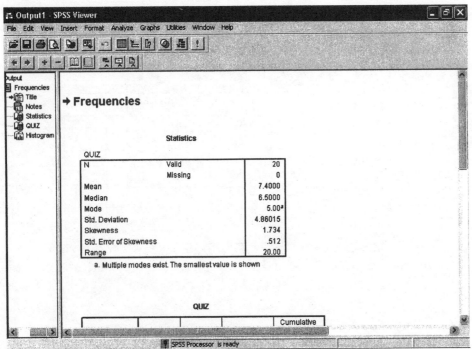

SF-4.7. Descriptive and graphical output for the frequency analysis.

42

Table 4.1 displays the descriptive statistics for the analysis. It also indicates the number of participants used for the analysis ($N = 20$) and notes that none of the cases were missing.

Table 4.1
Descriptive Statistics

Statistics

QUIZ

N	Valid	20
	Missing	0
Mean		7.4000
Median		6.5000
Mode		5.00[a]
Std. Deviation		4.86015
Skewness		1.734
Std. Error of Skewness		.512
Range		20.00

a. Multiple modes exist. The smallest value is shown

Table 4.2 displays the actual frequency counts for the raw data. This table also provides information regarding the percents, valid percents, and cumulative percents for the data.

Table 4.2
Frequency Counts for the Raw Data

QUIZ

		Frequency	Percent	Valid Percent	Cumulative Percent
Valid	.00	1	5.0	5.0	5.0
	3.00	1	5.0	5.0	10.0
	4.00	2	10.0	10.0	20.0
	5.00	3	15.0	15.0	35.0
	6.00	3	15.0	15.0	50.0
	7.00	3	15.0	15.0	65.0
	8.00	3	15.0	15.0	80.0
	9.00	1	5.0	5.0	85.0
	10.00	1	5.0	5.0	90.0
	20.00	2	10.0	10.0	100.0
	Total	20	100.0	100.0	

Lastly, Figure 4.1 graphically displays the histogram. At the top left portion of the histogram SPSS reminds you of the variable name in case you have conducted numerous analyses. The Y axis or ordinate, titled "Frequency," displays the frequencies in which numbers occurred in the data set. The X axis or abscissa, titled "QUIZ" displays the quiz scores.

Figure 4.1
Histogram

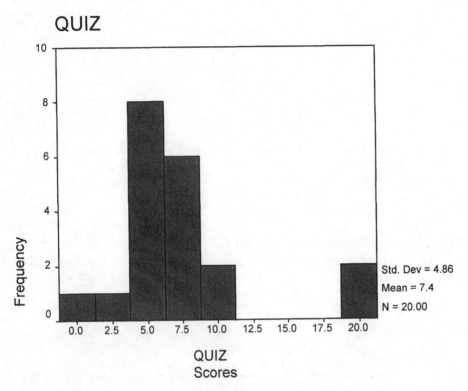

7. **Interpreting the Results**.

Describe the shape of the histogram distribution. Are there outliers in this data set?

- This distribution is bimodal because it has two prominent categories (scores for 5.0 and 7.5) or high points. Technically, the frequencies in the two high points (i.e., 5.0 and 7.5) must be exactly the same for this distribution to be called a bimodal distribution. Despite these restrictions, researchers rarely demand exact equivalence of the high points in bimodal distributions. Usually a description of the general shape of the distribution appears to serve the purpose.

- The distribution appears to be positively skewed (skewness = + 1.73, see Table 4.1) given that there are a cluster of lower quiz scores and that the tail of the distribution is on the right hand side.

- The distribution is also peaked indicating a smaller amount of score variability. If a normal distribution were applied to our distribution (as shown in Figure 4.3) it becomes a little easier to recognize the positive skew of the distribution.

- The two test scores of 20 would likely be considered outliers given their distance from the other test scores.

Figure 4.3
A Normal Distribution Applied to Our Histogram

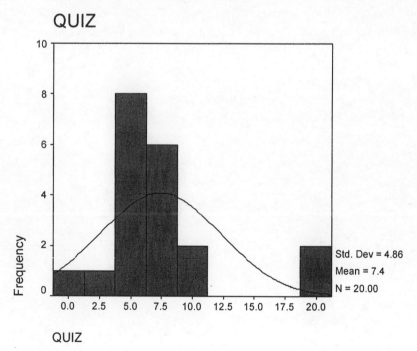

<u>Describe the measures of central tendency and explain which one best describes the data</u>?

- **Mode**: the most frequently occurring numbers for this data set were 5, 6, 7, and 8 given that they all occurred three times.

- **Median**: after rank ordering the 20 quiz scores we find that we have an even number of scores, so the median will lie half way between a score of 6 and 7. Thus, the median would be 6.5.

- **Mean**: the arithmetic average of the scores was 7.4.

- In this example the mode, median, and mean appear to closely describe the typical quiz score. The mode may not be the best choice given that the distribution is multimodal with a range of 5 to 8. The mode is typically used less often because it is less stable than the mean and median. The mean may also not be the best choice in this example because it is inflated given that there were two scores of 20 on the quiz and because 65 percent of the quiz scores were below the mean. In distributions that are approximately symmetrical, either the mean or the median will represent the average or typical score equally well, but with skewed distributions, as in this example, the median generally seems preferable because it is affected less by a few cases (two scores of 20) out in the long tail.

Answers to the multiple choice, matching, and true/false items

Multiple Choice	Matching	True and False
1. D	1. E	1. T
2. B	2. F	2. T
3. B	3. D	3. F
4. A	4. H	4. T
5. A	5. J	5. F
6. D	6. I	6. T
7. C	7. A	7. F
8. B	8. G	8. T
9. A	9. C	9. F
10. C	10. B	10. T
11. B		
12. D		
13. B		
14. A		
15. D		
16. C		
17. A		
18. B		
19. D		
20. B		

Chapter 5

The Basics of Experimentation I: Variables and Control

Learning Objectives

By the end of this chapter you should be able to:

1. Summarize why researchers operationally define variables.

2. Identify and describe independent variables (IVs), dependent variables (DVs), and extraneous variables.

3. List and describe several general categories of IVs.

4. Explain how extraneous variables can unintentionally influence the results of an experiment.

5. Distinguish between extraneous and nuisance variables.

6. List and summarize the basic techniques used for controlling extraneous and nuisance variables.

7. Describe the characteristics of a good DV.

8. Discuss the process for selecting and measuring DVs.

9. List and describe several factors that can threaten the internal validity of an experiment.

Expanded Outline

I. The Nature of Variables
    A Variable

II. Operationally Defining Variables
    Operational Definitions

III. Independent Variables (IV)
    Types of IVs
        Physiological
        Experience
        Stimulus or Environmental
        Participant Characteristics

48

IV.   Extraneous Variables (Confounders)

        Confounded

V.   Dependent Variables (DV)
            Selecting the DV
            Recording or Measuring the DV
                Correctness
                Rate or Frequency
                Degree or Amount
                Latency or Duration
            Should You Record More than One DV?

            Characteristics of a Good DV
                Valid
                Reliable

VI.   Nuisance Variables

VII.   Review Summary

VIII.   Check Your Progress

IX.   Controlling Extraneous Variables
            Basic Control Techniques
                Randomization

                Elimination
                Constancy

                Balancing

                Counterbalancing

                    Within-Subject Counterbalancing
                    Within-Group Counterbalancing
                    Complete Counterbalancing
                    Incomplete Counterbalancing
                    Sequence or Order Effects
                    Carryover Effects
                    Differential Carryover

Practice Exam

Multiple Choice

Identify the letter of the choice that best completes the statement or answers the question.

_____ 1. In 1920 physicist Percy W. Bridgman proposed that researchers should define their variables in terms of the operations needed to produce them. What was Bridgman referring to in this statement?

    a. identifying IVs   c. operationally defining variables
    b. measuring DVs   d. eliminating extraneous variables

_____ 2. All of the following are types of IVs <u>except</u>

    a. physiological.   c. experience.
    b. operational.   d. stimulus or environmental.

_____ 3. Steve is conducting a study about first name stereotypes and perceived vocational roles in business settings. He is using the same resume, but has changed the names of the two female job applicants so that the first applicant is named Angel and the second applicant is named Deborah. He has manipulated the two names to determine if they will have an effect on employers' selection of a job candidate. What type of IV is he using?

    a. stimulus   c. experience
    b. physiological   d. participant

_____ 4. Researchers using a(n) _____ IV assign participants to conditions that alter or change their normal biological state.

    a. stimulus   c. experience
    b. physiological   d. participant

_____ 5. When researchers can attribute the results of an experiment to an IV or to an extraneous variable, the experiment is considered to be

    a. successful.   c. meaningful.
    b. confounded.   d. none of the above

_____ 6. An independent variable is a(n)

    a. manipulated variable.   c. unwanted factor affecting the study.
    b. measured variable.   d. none of the above

____ 7. A dependent variable is a(n)

    a. manipulated variable.   c. unwanted factor affecting the study.
    b. measured variable.   d. none of the above

____ 8. An extraneous variable is a(n)

    a. manipulated variable.   c. unwanted factor affecting the study.
    b. measured variable.   d. none of the above

____ 9. A baseball manager conducted a study by changing the color of his team's uniform to determine if the new color would cause his team to win more games. What is the DV in this study?

    a. uniform color   c. number of wins
    b. sex of the manager   d. age of the players

____ 10. Measuring DVs may be accomplished by recording

    a. correctness.   d. latency or duration.
    b. rate or frequency.   e. all of the above
    c. degree or amount.

____ 11. Researchers at Wheaton College were interested in the relation between a participant's mood and his or her reaction time. What type of DV were they measuring?

    a. correctness   c. degree or amount
    b. rate or frequency   d. latency or duration

____ 12. A good DV should be

    a. reliable.   d. both b and c
    b. valid.   e. all of the above
    c. directly related to the IV.

____ 13. Professor Parker suspects that a nuisance variable was present in his study. What factors could he investigate to help him isolate the presence of a nuisance variable or variables?

    a. participant characteristics   c. both a and b
    b. unintended affects of the   d. none of the above
        experimental situation

14. What control technique guarantees that each participant has an equal chance of being assigned to any group in the experiment?

   a. constancy   d. balancing
   b. randomization   e. counterbalancing
   c. elimination

15. Annie wants to conduct her research project over a five day period to obtain as many participants as possible. She has decided to collect the data in the same room at the same time each day. What type of control technique is she using?

   a. constancy   d. balancing
   b. randomization   e. counterbalancing
   c. elimination

16. In a study, the experimental group is scheduled to be run by an African-American researcher and the control group by a Caucasian researcher. Since the researchers do not want ethnicity to be a confounder in their study, they decide to each run half of the experimental group, and half of the control group. Which control technique will the researchers likely use?

   a. constancy   d. balancing
   b. randomization   e. counterbalancing
   c. elimination

17. Which of the following control techniques is best suited to address order or sequence problems?

   a. constancy   d. balancing
   b. randomization   e. counterbalancing
   c. elimination

18. _____ attempts to control sequence effects within each participant.

   a. Balancing   c. Within-group counterbalancing
   b. Within-subject counterbalancing d. Control

19. Tony has been asked to conduct a pilot study of two new mouthwash flavors. He is concerned about controlling for sequence effects. He decides to randomly assign half of the participants to taste flavor A followed by flavor B; whereas the remaining participants receive flavor B followed by flavor A. What control technique is he using?

   a. balancing
   b. within-subject counterbalancing
   c. within-group counterbalancing
   d. constancy

20. The formula for *n*! or *n* factorial is commonly associated with

   a. randomization.
   b. complete counterbalancing.
   c. incomplete counterbalancing.
   d. balancing.

21. Tony (please refer back to question 19) has decided to increase the amount of mouthwash flavors in his study from two to three. How many participants (at a minimum) would be required to completely counterbalance his experiment? (Hint: use the *n*! formula)

   a. 6
   b. 8
   c. 10
   d. 12

22. Which of the following is not considered to be a problem associated with incomplete counterbalancing?

   a. each treatment does not appear an equal number of times at each testing session
   b. each treatment does not precede and follow each of the other treatments an equal number of times
   c. each participant receives each treatment an equal number of times
   d. none of the above

23. A(n) _____ is present when the effects of one treatment persists and influences the participant responses to future treatment.

   a. carryover effect
   b. randomization effect
   c. elimination effect
   d. incomplete counterbalancing effect

24. Six judges were asked to rate three types of wine presented in different sequential orders. Researchers found that overall the judge's rating of wine B was dependent on whether they had tasted wine C first. What type of phenomenon is occurring in this example?

   a. a carryover effect
   b. a differential carryover effect
   c. a constancy effect
   d. a sequence or order effect

25. Sam is concerned that the IV he used in his study may not have been responsible for the changes he observed in his DV. What type of validity is he concerned about?

    a. discriminant   c. external
    b. concurrent   d. internal

26. _____ refers to systematic time-related changes that often occur in experiments extending over time and ultimately threatens the internal validity of the IV.

    a. History   c. Testing
    b. Maturation   d. Statistical regression

27. Doctor Davis is midway through the data collection process when suddenly a fire alarm rings and distracts his participants. He is concerned that this distraction may cause the current participants to respond differently from the participants studied earlier in the week. Which threat to internal validity should he primarily be concerned about?

    a. history   c. testing
    b. maturation   d. selection

28. _____ is a threat to internal validity that often occurs because the process of measuring the DV actually causes a change in the DV.

    a. History   c. Instrumentation
    b. Maturation   d. Testing

29. Double blind research designs can be used to control for _____ problems associated with the researcher's observations and eventually the coding of the data.

    a. reactivity   c. maturation
    b. instrumentation   d. history

30. Which of the following threats to internal validity typically occur when low scoring participants improve or when high scoring participants perform worse on a second test administration due primarily to statistical reasons.

    a. selection   c. statistical regression
    b. mortality   d. history

31. Sadie hastily chooses participants to complete her class research project that is due soon. Her advisor warns that she may have selected participants in such a way that the comparison groups are not equal before the experiment. What internal validity threat is her advisor likely concerned about in this situation?

a. selection
b. mortality
c. statistical regression
d. history

32. Researchers often begin studying twins at an early age and use longitudinal studies to investigate their physical and cognitive growth and development. However, researchers struggle to keep the original twin comparison groups in tact because of experimental dropouts. What type of internal validity threat does this illustrate?

a. selection
b. mortality
c. maturation
d. history

33. Which threat to internal validity may occur when the selected participant groups demonstrate differences on another variable that varies systematically within the participant groups?

a. maturation
b. history
c. interactions with selection
d. statistical regression

34. You and your roommate choose to participate in a psychological research study, but cannot attend the same experimental sessions. Your roommate attends the first session and excitedly calls you to discuss the details of the study. This example illustrates _____.

a. interaction with selection
b. differential carryover
c. sequence/order effects
d. diffusion or imitation of treatments

Matching

a. validity

b. maturation

c. elimination

d. variable

e. reliability

f. DV

g. within-group counterbalancing

h. sex

i. constancy

j. mortality

k. uncontrolled variables

l. within-subject counterbalancing

m. nuisance variable

_a_ 1. measuring what you intend to measure

_f_ 2. measures a response or behavior

_k_ 3. extraneous variables

_g_ 4. presenting different treatment sequences to different participants

_e_ 5. consistency in measurement

_j_ 6. related to participants withdrawing from an experiment

_c_ 7. an attempt to remove all extraneous variables

_l_ 8. presenting different treatment sequences to the same participant

_b_ 9. changes in participants that occur over time during the experiment

_m_ 10. increases the score variability within groups

_h_ 11. a participant characteristic

_d_ 12. an event or behavior that can assume two or more values

_i_ 13. reducing an extraneous variable to a single value

True/False

Indicate whether the sentence or statement is true or false.

_____   1.   A variable is an event or behavior that can assume at least two values.

_____   2.   Participant characteristics or IVs such as sex and age are best viewed as classification and <u>not</u> manipulation variables.

_____   3.   It is relatively easy to identify extraneous variables.

_____   4.   It is possible to record more than one DV in an experiment.

_____   5.   Nuisance variables decrease the spread of scores within a distribution.

_____   6.   Elimination is the most widely used control technique in research.

_____   7.   Counterbalancing is used when a sequence or order effect must be controlled.

_____   8.   Counterbalancing effectively protects against differential carryover.

_____   9.   Randomization is the most widely used control procedure.

_____   10.   Constancy involves the complete removal of the extraneous variable.

_____   11.   A participant's right to withdraw from an experiment at any point without penalty influences mortality rates and threatens internal validity.

_____   12.   Selecting participant groups in such a way that they are not equal before the experiment threatens internal validity.

Short Answer

1.   List and describe the types of IVs. Which of these IVs is not considered to be a true IV? Explain your answer.

2.   Discuss why the presence of an extraneous variable can be devastating to research.

3.   Explain the reason(s) for researchers recording more than one DV while conducting a study. What factors should researchers consider before making this type of decision?

4.   Distinguish between extraneous and nuisance variables.

5.   List and describe the five basic control techniques in research. Discuss when and why researchers should use these techniques.

6. Summarize why randomization is one of the most frequently used control techniques in research. Discuss some of the problems associated with this technique.

7. Explain why balancing represents a logical extension of the control technique known as constancy.

8. Describe the differences between within-subject and within-group counterbalancing.

9. Discuss a major disadvantage related to using within-subject counterbalancing.

10. Distinguish between complete and incomplete counterbalancing.

11. Summarize the differences between carryover and differential carryover effects.

12. Describe Cambell's (1957) rationale for advocating the use of non-reactive measures in psychological research. Discuss how reactive measures can influence the outcome of a study.

13. Contrast and compare internal and external validity. Provide several examples of concepts that may threaten the internal validity of a study.

14. Define statistical regression and discuss why it is related to internal validity.

15. Explain why a participant's right to withdraw from an experiment at any time is related to the internal validity threat known as mortality.

16. Provide an example of a situation that resembles a diffusion or imitation of treatment(s).

Answers to the multiple choice, matching, and true/false items

Multiple Choice		Matching		True and False	
1.	C	1.	A	1.	T
2.	B	2.	F	2.	T
3.	A	3.	K	3.	F
4.	B	4.	G	4.	T
5.	B	5.	E	5.	F
6.	A	6.	J	6.	F
7.	B	7.	C	7.	T
8.	C	8.	L	8.	F
9.	C	9.	B	9.	T
10.	E	10.	M	10.	F
11.	D	11.	H	11.	T
12.	E	12.	D	12.	T
13.	C	13.	I		
14.	B				
15.	A				
16.	D				
17.	E				
18.	B				
19.	C				
20.	B				
21.	A				
22.	C				
23.	A				
24.	B				
25.	D				
26.	B				
27.	A				
28.	D				
29.	B				
30.	C				
31.	A				
32.	B				
33.	C				
34.	D				

Chapter 6

Summarizing and Comparing Data: Measures of Variation, Distribution of Means and the Standard Error of the Mean, and *z*-Scores

Learning Objectives

By the end of this chapter you should be able to:

1. List and describe the steps for calculating the range, variance, and standard deviation.

2. Summarize and interpret data using the measures of variability.

3. Explain the relationships between the measures of variability and a normal distribution.

4. Describe and interpret the three types of kurtotic distributions.

5. Define and discuss the purpose of the standard error of the mean.

6. Discuss the characteristics of a *z*-distribution.

7. Calculate and interpret *z*-scores.

8. Convert *z*-scores back to raw scores.

9. Describe the steps involved with calculating 68, 95, and 99 percent confidence intervals.

10. Interpret 68, 95, and 99 percent confidence intervals.

By the end of this chapter you should learn when and how to use the following formulas:

1. <u>Range</u> = (Highest score – Lowest score)

2. <u>Deviation score formulas for calculating the sample variance</u>

$$S^2 = \frac{\Sigma(X-M)^2}{N} \qquad \textbf{OR} \qquad S^2 = \frac{\Sigma(x)^2}{N}$$

Where:

S^2	=	sample variance
Σ	=	sum
X	=	raw score
M	=	mean of all scores
x	=	deviation score
N	=	total number of participant scores

3. Raw score formula for calculating the sample variance

$$S^2 = \frac{\Sigma X^2 - \frac{(\Sigma X)^2}{N}}{N}$$

Where:

S^2	=	sample variance
Σ	=	sum
X	=	raw score
N	=	total number of participant scores

4. Computational formula for calculating the sample standard deviation

$$SD = \sqrt{\frac{\Sigma(X-M)^2}{N}}$$

Where:

Σ	=	sum
X	=	raw score
M	=	mean of all scores
N	=	total number of participant scores

5. The formula for calculating a z-score

$$z = \frac{X-M}{SD}$$

Where:

z	=	z-score
X	=	raw score
M	=	mean score
SD	=	standard deviation

6. <u>The formula for transforming a z-score to a raw score</u>

$$X = (z)(SD) + M$$

Where:

X	=	raw score
z	=	z-score
SD	=	standard deviation
M	=	mean score

Expanded Outline

I.   Measures of Variability
        Variability
        Range
        Variance
        Deviation Score

        Calculating and Interpreting the Standard Deviation
                Normal Distribution

                Kurtosis
                        Leptokurtic
                        Mesokurtic
                        Platykurtic

II.   Review Summary

III. Check Your Progress

IV.   Distribution of Means and the Standard Error of the Mean

        Distribution of Means
        Standard Error of the Mean

V.   z-Scores

        Confidence Intervals
                Interval Estimates
                        68, 95, and 99 percent
                Parameter

VI.   Review Summary

VII.   Check Your Progress

VIII.   Exercises

IX.   Looking Ahead

X.   Key terms

Practice Exam

Multiple Choice

Identify the letter of the choice that best completes the statement or answers the question.

_____ 1.   Statistical procedures that allow researchers to organize, summarize, and describe
            the characteristics of a sample of scores are called _____ procedures.

            a. independent   c. descriptive
            b. dependent   d. inferential

_____ 2.   What is the range of the following set of numbers: 7, 10, 12, and 16?

            a. 7   c. 9
            b. 8   d. 10

_____ 3.   The _____ is a single number that represents the total amount of variability.

            a. variance   c. range
            b. standard deviation   d. mean

_____ 4. The average squared deviation is called the

    a. standard deviation.   c. sum of squares.
    b. average deviation.   d. variance.

_____ 5. The square root of the average squared deviation is called the

    a. average deviation.   c. variance.
    b. standard deviation.   d. sum of squares.

_____ 6. What is the variance and standard deviation for the following scores?

1, 5, 5, 9

    a. 4, 2   c. 8, 2.82
    b. 5, 2.23   d. 16, 4

_____ 7. The standard deviation (SD) is preferred over the variance as a measure of variability because the

    a. variance is an inaccurate measure.   c. SD is more conservative.
    b. variance doesn't account for extreme scores.   d. SD is easier to interpret.

_____ 8. Describing a distribution involves explaining the

    a. shape, central tendency, and skewness.   c. kurtosis, shape, and variation.
    b. shape, central tendency, and variation.   d. kurtosis, skewness, and variation.

_____ 9. The majority of scores in a _____ distribution cluster around the measure of central tendency with fewer and fewer scores occurring as we move away from it.

    a. normal   c. negatively skewed
    b. positively skewed   d. none of the above

_____ 10. What percentage of the scores lie within the first SD above and below the mean in a normal distribution (i.e., -1 SD to +1 SD)?

    a. 13.59   c. 47.72
    b. 34.13   d. 68.26

_____ 11. _____ refers to the degree of peakedness (i.e., flat or tall) in a symmetrical distribution.

    a. Variance   c. Kurtosis
    b. Skewness   d. Standard deviation

_____ 12.   A leptokurtic distribution can be described as

        a. nearly normal.   c. broad and flat.
        b. tall and peaked.   d. none of the above

_____ 13.   Which distribution shape will likely have the largest standard deviation?

        a. platykurtic   c. mesokurtic
        b. leptokurtic   d. normal

_____ 14.   Consider a normal distribution with a mean of 100 and a standard deviation of 10. What percentage of the scores will fall between 80 and 120?

        a. 13.59   c. 68.26
        b. 34.13   d. 95.44

_____ 15.   In a normal distribution, you know that the mean of the scores is 50 and the SD is 5. Where would a score of 40 fall?

        a. 1 SD above the mean   c. 2 SDs above the mean
        b. 1 SD below the mean   d. 2 SDs below the mean

_____ 16.   A distribution of means will approximate a normal distribution if the population is a normal distribution <u>or</u> if each randomly selected sample contains at least _____ scores?

        a. 30   c. 90
        b. 60   d. 120

_____ 17.   The SD of a distribution of means is referred to as the

        a. variance.   c. standard error of the mean.
        b. square root.   d. standard error of the median.

_____ 18.   When the sample size of a distribution increases, the standard error of the mean

        a. increases.   c. decreases.
        b. stays the same.   d. varies randomly.

_____ 19.   A _z_-score indicates how much a score from a distribution deviates from the

        a. mean in raw score units.   c. mean in SD units.
        b. median in percentage units. d. other scores in percentage units.

20. The formula for transforming a raw score in a sample to a z-score is mathematically expressed as:

   a. M - X / SD   c. M + X / SD
   b. X - M / SD   d. X + M / SD

21. Peggy received a z-score of 1.5. If the mean of the distribution of test scores was 71 and the standard deviation was 10, what is Peggy's raw test score?

   a. 66   c. 86
   b. 76   d. 96

22. Sunshine is administered a questionnaire to measure her depression. A few weeks later she learned that she received a score of 72. What is her z-score if the mean was 60 and the standard deviation was 10?

   a. 0.2   c. 1.2
   b. 1.0   d. 2.0

23. Using a z-table, approximately what percentage of people scored at or below Sunshine's score (refer back to question 22)?

   a. 12   c. 68
   b. 24   d. 88

24. Positive z-scores are based on raw scores that are _____ the mean; while negative z-scores are based on raw scores that are _____ the mean.

   a. equal to, not equal to   c. above, below
   b. below, above   d. none of the above

25. A range of values around a sample mean, possibly containing the population mean, is referred to as a(n)

   a. central limit theorem.   c. standard error.
   b. confidence interval.   d. average deviation.

26. Which of the following confidence intervals is more likely to include the population characteristic of interest?

   a. 34   c. 95
   b. 68   d. 99

_____ 27.   Which of the following *z*-scores correspond with 95 and 99 percent confidence
            intervals respectively?

    a. 1.00 and 1.96   c. 2.57 and 3.00
    b. 1.96 and 2.57   d. 3.00 and 3.57

_____ 28.   A teacher reported a mean of 67 and a standard deviation of 10 for her history
            test. What is the 95 percent confidence interval for this data?

    a. 35.9 to 85.6   c. 54.5 to 90.7
    b. 47.4 to 86.6   d. 67.0 to 97.0

_____ 29.   What is the 99 percent confidence interval for the data in question 28?

    a. 25.6 to 99.6   c. 41.3 to 92.7
    b. 33.7 to 95.0   d. 50.8 to 88.9

Matching Terms and Concepts

a. N

b. z-score of 2.57

c. S^2

d. z

e. μ_m

f. $\Sigma(X-M)^2 / N$

g. ΣX

h. M

i. mesokurtic

j. Σ

k. leptokurtic

l. platykurtic

m. X

n. $\sqrt{S^2}$

o. $(X - M) = x$

p. z-score of 1.96

q. μ

r. $\Sigma(X-M)^2$

s. x^2

_____ 1. used to calculate the variance

_____ 2. a 99 percent confidence interval is equivalent to a

_____ 3. the formula for a deviation score

_____ 4. resembles a normal distribution

_____ 5. symbol for the variance

_____ 6. a term for a broad and flat distribution

_____ 7. a 95 percent confidence interval is equivalent to a

_____ 8. symbol indicating the scores should be summed

_____ 9. the sum of all scores in the distribution of the variable X

_____ 10. symbol for the mean of a population

_____ 11. method for calculating the SD

_____ 12. symbol for the number of participant scores or observations in a distribution

_____ 13. symbol for each score in the distribution of the variable X

_____ 14. symbol for the mean of a distribution of means

_____ 15. formula for the sum of squared deviations from the mean

_____ 16. symbol for the squared deviation score

_____ 17. symbol for a z-score

_____ 18. symbol for the mean

_____ 19. term for a tall and peaked distribution

Matching the Components of a Formula

Match each symbol with its corresponding term for each formula

A. <u>Deviation score formulas for calculating the sample variance</u>

$$S^2 = \frac{\Sigma(X-M)^2}{N} \qquad \textbf{OR} \qquad S^2 = \frac{\Sigma(x)^2}{N}$$

20. sample variance _____

21. sum _____

22. deviation score _____

23. number of scores _____

24. raw score _____

25. mean of all scores _____

B. <u>Computational formula for calculating the sample standard deviation</u>

$$SD = \sqrt{\frac{\Sigma(X-M)^2}{N}}$$

26. sum _____

27. standard deviation _____

28. number of scores _____

29. raw score _____

30. mean of all scores _____

C. The formula for calculating a *z*-score

$$z = \frac{X - M}{SD}$$

31. *z*-score   _____

32. standard deviation   _____

33. raw score   _____

34. mean score   _____

D. The formula for transforming a z-score to a raw score

$$X = (z)(SD) + M$$

35. *z*-score   _____

36. standard deviation   _____

37. raw score   _____

38. mean score   _____

True/False

Indicate whether the sentence or statement is true or false.

____ 1. A standard deviation is a measure of central tendency that is frequently reported by psychologists.

____ 2. In order to obtain the standard deviation, you must first calculate the variance.

____ 3. The sum of the deviation scores will <u>never</u> equal zero because the deviation scores are evenly distributed above and below the mean.

____ 4. *N* - 1 is used when you want to estimate the values of a larger group of people (i.e., the population).

____ 5. A mesokurtic distribution is characterized as broad and flat.

____ 6. A normal distribution closely resembles a leptokurtic distribution.

_____ 7. A distribution of means is based on scores from individual participants.

_____ 8. The standard error of the mean describes the amount of variability in the population from which you obtained your sample.

_____ 9. A *z*-score indicates the amount that a score deviates from the mean when measured in standard deviation units.

_____ 10. A *z*-score <u>always</u> has a mean of zero and a standard deviation of one.

_____ 11. Researchers construct interval estimates to identify the range or interval that is likely to include sample characteristics.

_____ 12. An interval estimate with a percentage associated with it is called a confidence interval.

Short Answer

1. You receive a score of 72 on your psychology exam. What does this number tell you? What additional information do you need to accurately describe your score?

2. Discuss why the range is considered an insufficient measure of variability?

3. Describe the steps required to calculate the variance and SD for a set of numbers using the raw score or computational formula.

4. Summarize the relationship between the SD and a normal distribution.

5. Describe and draw the three types of kurtotic distributions. Explain why the SDs vary for each distribution.

6. Explain why it is important to understand the variability of means in a population.

7. Define the standard normal curve model? Discuss why it is used? What criterion should be met for the model to give an accurate description of a sample?

8. Describe the characteristics of a *z*-distribution and how it is used to compare different distributions.

9. List and describe the steps involved with calculating a confidence interval. Explain the difference(s) between 68, 95, and 99 percent confidence intervals and how they are interpreted.

Answers to the multiple choice, matching, and true/false items

Multiple Choice		Matching		True and False	
1.	C	1.	F	1.	F
2.	C	2.	B	2.	T
3.	A	3.	O	3.	F
4.	D	4.	I	4.	T
5.	B	5.	C	5.	F
6.	C	6.	L	6.	F
7.	D	7.	P	7.	F
8.	B	8.	J	8.	T
9.	A	9.	G	9.	T
10.	D	10.	Q	10.	T
11.	C	11.	N	11.	F
12.	B	12.	A	12.	T
13.	A	13.	M		
14.	D	14.	E		
15.	D	15.	R		
16.	A	16.	S		
17.	C	17.	D		
18.	C	18.	H		
19.	C	19.	K		
20.	B	20.	S^2		
21.	C	21.	Σ		
22.	C	22.	x		
23.	D	23.	N		
24.	C	24.	X		
25.	B	25.	M		
26.	D	26.	Σ		
27.	B	27.	SD		
28.	B	28.	N		
29.	C	29.	X		
		30.	M		
		31.	z		
		32.	SD		
		33.	X		
		34.	M		
		35.	z		
		36.	SD		
		37.	X		
		38.	M		

Chapter 7

The Basics of Experimentation II: Final Considerations, Unanticipated Influences, and Cross-Cultural Issues

Learning Objectives

By the end of this chapter you should be able to:

1. Summarize the importance of the type and number of participants used in research.

2. Describe a wide variety of apparatus used to present the IV(s) and record the DV.

3. Discuss several experimenter characteristics that may influence participant responses.

4. List and describe the most common procedures for controlling general experimenter effects or characteristics during the research process.

5. Recognize that a participant's perception of the research project can operate as an extraneous variable.

6. List and summarize the most common procedures for controlling participant effects during the research process.

Expanded Outline

    I.   Participants
               Type of Participants
                            Precedent

                            Availability
                            Type of Research Project
               Number of Participants
                            Finances
                            Time
                            Availability

                            Power

II.   Apparatus
            IV Presentation
            DV Recording

III.   Review Summary

IV.   Check Your Progress

V.   The Experimenter as an Extraneous Variable
            Experimenter Characteristics

            Experimenter Expectancies

                  Rosenthal Effects
            Controlling Experimenter Effects
                  Physiological and Psychological Effects
                  Experimenter Expectancies
                        Single-Blind Experiment

VI.   Participant Perceptions as Extraneous Variables
            Demand Characteristics and Good Participants
                  Demand Characteristics
                  Good Participant Effect

            Response Bias
                  Yea-Saying
                      Yea-Sayers
                      Nay-Sayers
                  Response Set
            Controlling Participant Effects
                  Demand Characteristics
                      Double-Blind Experiments

                  Yea-Saying
                  Response Set

VII.   Review Summary

VIII.   Check Your Progress

Practice Exam

Multiple Choice

Identify the letter of the choice that best completes the statement or answers the question.

_____ 1. Which of the following guidelines should be considered when selecting participants for a research study?

     a. the nature of the problem   c. precedence
     b. availability   d. all of the above

_____ 2. Sam is reviewing previous research studies to investigate which type of participant he will use to conduct his study. Which guideline is he considering?

     a. the nature of the problem   c. precedence
     b. availability   d. none of the above

_____ 3. The continued use of a particular species or participant can limit the _____ of the information gathered in research.

     a. generalizability   c. precedence
     b. likelihood of success   d. power

_____ 4. In your psychology class you are required to participate in a research study. This requirement ensures the _____ of the participants.

     a. generalizability   c. appropriateness
     b. availability   d. randomness

_____ 5. As the within-group variability of the participant sample increases, the number of participants needed to conduct the study

     a. stays the same.   c. increases.
     b. decreases.   d. becomes meaningless.

_____ 6. Alex and Steve decide to conduct similar research projects to compare their results. However, Alex predicts that his participant sample will exhibit less within-group variability. What concept best describes Alex's participant sample?

     a. biased   c. heterogeneous
     b. unbiased   d. homogeneous

_____ 7. Which of the following terms is related to the likelihood or probability that the statistical test we use to analyze our data will be significant?

a. power
b. response set
c. generalizability
d. ethnocentrism

_____ 8. Researchers manipulate the _____ variable and record or measure the _____ variable.

a. extraneous, independent
b. independent, dependent
c. nuisance, dependent
d. dependent, independent

_____ 9. Tiger would like to investigate the effects of frustration on a golfer's performance, but he does not know how to measure this effect. Tiger is struggling to

a. present the IV.
b. record the IV.
c. present the DV.
d. record the DV.

_____ 10. In 1977, Robert Rosenthal demonstrated that male researchers were more friendly to their participants than were female researchers. What concept did he demonstrate?

a. psychological characteristics
b. physiological characteristics
c. experimenter expectations
d. none of the above

_____ 11. Researchers may control experimenter expectancies by using

a. objective and concrete instructions.
b. single-blind experiments.
c. instrumentation and automation.
d. all of the above
e. none of the above

_____ 12. Sally has diligently constructed her experiment so that she and the participants are unaware of which treatment the participants are receiving. Which experimental control method is she using?

a. single-blind
b. double-blind
c. mixed-blind
d. response sets

____ 13. Pete is conducting a study about nightmares and the participants are required to spend the night in the research lab. However, he recognizes that the participants are acting abnormal because of their expectancies about the research project. This type of participant behavior illustrates the effect of

    a. a blind experiment.   c. demand characteristics.
    b. environmental stimuli.   d. physiological variables.

____ 14. The desire to act as the participants believe the experimenter wants them to act is known as a(n) _____ effect.

    a. good participant   c. ethnocentric
    b. biased   d. nay-sayer

____ 15. Participants who attempt to respond in socially desirable ways are exhibiting

    a. nay-saying.   c. response bias.
    b. the Rosenthal effect.   d. demand characteristics.

____ 16. Sadie is diligently reviewing her research questions to determine if they will invoke socially desirable responses. She has also examined the nature of her experimental environment to avoid the presence of undesirable cues. What is she attempting to safeguard against?

    a. nay-sayers   c. Rosenthal effects
    b. response sets   d. balancing

____ 17. The goal of _____ is to determine whether research results and psychological phenomena are universal or specific to the culture in which they were reported.

    a. social psychology   c. sociology
    b. educational psychology   d. cross-cultural psychology

____ 18. An _____ finding is linked to a specific culture; whereas an _____ finding occurs across cultures.

    a. emic, ethnocentric   c. etic, emic
    b. emic, etic   d. ethnocentric, etic

19. The results of Javier's study on cultural learning styles revealed that students in the United States preferred to study independently; whereas students in China preferred to study in groups. This example best describes a(n) _____ finding.

   a. ethnocentric
   b. etic
   c. emic
   d. biased

20. The perception that other cultures are an extension of their own culture best describes

   a. ethnocentrism.
   b. an etic.
   c. an emic.
   d. bias.

21. Yanyan is conducting cross-cultural research and realizes that she may not be able to present the same stimulus items via computer to both cultures. Consequently, she develops an alternative booklet form to present the stimulus items. She is having difficulty with

   a. sampling procedures.
   b. selecting an IV.
   c. selecting a DV.
   d. data analysis.

22. The tendency of a particular culture to respond in a certain manner best describes a(n)

   a. etic.
   b. emic.
   c. Rosenthal effect.
   d. cultural response set.

Matching

a. culture response set   g. power

b. ethnocentrism   h. yea-sayers

c. etic   i. precedence

d. culture   j. Rosenthal effect

e. nay-sayers   k. emic

f. increased participant heterogeneity   l. increased participant homogeneity

_____ 1.   small within-group variability

_a__ 2.   tendency of one culture to respond in a certain manner

_____ 3.   large within-group variability

_g__ 4.   probability that a statistical test will be significant

_h__ 5.   participants who tend to answer yes

_b__ 6.   when cultures are viewed as an extension of one's culture

_d__ 7.   lasting values, attitudes, and behaviors that a group of people share

_c__ 8.   culture specific finding

_e__ 9.   participants who tend to answer no

_k__ 10. a finding that is similar in different cultures

_j__ 11. describes experimenter expectancies

_i__ 12. using previous research to select participants

True/False

Indicate whether the sentence or statement is true or false.

_____ 1. Participants exhibiting smaller amounts of within-group variability are considered to be homogenous.

_____ 2. The variability of scores makes it less difficult to recognize absolute differences between participant groups.

_____ 3. Power is related to the likelihood that a statistical test will be significant.

_____ 4. Single-blind experiments are used to minimize experimenter expectancies.

_____ 5. The desire to cooperate and act as the participants believe the experimenter wants them to act is called ethnocentrism.

_____ 6. Individuals can be of the same race or nationality and not share the same culture.

_____ 7. The number of emics is considerably greater than the number of etics.

_____ 8. Culture influences all aspects of the research process.

_____ 9. A culture specific finding is known as an etic finding.

_____ 10. A cultural-response set often describes the tendency of a particular culture to respond in a certain manner.

Short Answer

1. Summarize why it is important to consider the type and the number of participants used in a research study.

2. Distinguish between homogeneity and heterogeneity. How are these concepts related to participant sample size?

3. Describe the differences between an IV and a DV. Provide an example illustrating how you might present an IV to your participants and describe the apparatus you will use. Then discuss your method for measuring the DV.

4. Define experimenter expectancy. Briefly list and describe the steps a researcher can take to control for experimenter expectancies.

5. Compare and contrast single-blind and double-blind studies. Explain when it is appropriate to use these control methods.

6. Discuss the roles that instrumentation and automation can play in helping researchers control for experimenter expectancies.

7.   Define and discuss the differences between emic and etic findings. Describe how these concepts may influence the results of an experiment.

8.   You have been asked to give a short presentation about cross-cultural research in your psychology class. What primary concepts and issues will you discuss? Be sure to emphasize how these concepts and issues can influence research results.

9.   Discuss the influences that culture may have on researchers when they are deciding which type of survey or questionnaire they will be using in their study.

10.   Explain why the goal of cross-cultural research is incompatible with ethnocentrism.

Answers to the multiple choice, matching, and true/false items

Multiple Choice		Matching		True and False	
1.	D	1.	L	1.	T
2.	C	2.	A	2.	F
3.	A	3.	F	3.	T
4.	B	4.	G	4.	T
5.	C	5.	H	5.	F
6.	D	6.	B	6.	T
7.	A	7.	D	7.	F
8.	B	8.	K	8.	T
9.	D	9.	E	9.	F
10.	B	10.	C	10.	T
11.	D	11.	J		
12.	B	12.	I		
13.	C				
14.	A				
15.	C				
16.	B				
17.	D				
18.	B				
19.	C				
20.	A				
21.	B				
22.	D				

Chapter 8

Correlation and Prediction

Learning Objectives

By the end of this chapter you should be able to:

1. Summarize the purposes, types, and limitations of correlations.

2. Construct scatterplots to graphically display correlations.

3. Calculate a correlation coefficient.

4. Interpret a correlation coefficient.

5. Evaluate the outcome of a study with respect to its statistical significance and effect size.

6. List and describe the components of a regression equation.

7. Construct and interpret a regression line using the regression equation.

By the end of this chapter you should learn when and how to use the following formulas:

1. Raw score formula for calculating the Pearson product-moment correlation coefficient (r).

$$r = \frac{N\Sigma XY - (\Sigma X)(\Sigma Y)}{\sqrt{\left[N\Sigma X^2 - (\Sigma X)^2\right]\left[N\Sigma Y^2 - (\Sigma Y)^2\right]}}$$

$$(1)$$

Where:

N	=	number of pairs of scores
X	=	variable X
Y	=	variable Y
ΣX	=	sum of scores for variable X
ΣY	=	sum of scores for variable Y
ΣXY	=	sum of the cross products for variables X and Y

2. Regression equation

Note: in order to calculate the regression equation you will first need to calculate the slope (b) using either formula 3 or 4 and the Y-intercept (a) using formula 5.

$$\hat{Y} = a + bX \tag{2}$$

Where:

$\hat{Y}$ = the "hat" indicates that we are making a prediction of a score on the Y variable (criterion variable)

a = the point at which the regression line intersects the Y axis

b = the slope of the regression line (the regression coefficient)

X = the predictor variable

Calculating the slope for the regression equation. This may be accomplished in two ways:

If you are given the correlation coefficient (r) and the standard deviations (SDs) for the X and Y variables then you can use the formula 3 to calculate the slope:

$$b = r_{xy} \frac{SD_y}{SD_x} \tag{3}$$

Where:

r_{xy} = the correlation coefficient for variables X and Y

SD_y = SD of variable Y

SD_x = SD of variable X

Formula 4 is an alternative formula for obtaining the slope of the linear equation. If you have calculated the XY correlation, then calculating b with this formula will be simple.

$$b = \frac{N(\Sigma XY) - (\Sigma X)(\Sigma Y)}{N(\Sigma X^2) - (\Sigma X)^2} \tag{4}$$

Where:

N = number of pairs of scores

X = variable X

Y = variable Y

ΣX = sum of scores for variable X

ΣY = sum of scores for variable Y

ΣXY = sum of the cross products for variables X and Y

<u>Calculating the Y-intercept for the regression equation.</u>

Once you have calculated the slope then you can calculate the <u>Y-intercept</u> by using formula 5.

$$a = M_{\hat{y}} - (b)(M_{\hat{x}})$$
(5)

Where:

$M_{\hat{y}}$	=	the mean of the Y variable
$M_{\hat{x}}$	=	the mean of the X variable
b	=	the regression coefficient

Expanded Outline

I. Correlation
   The Nature of Correlation
   The Scatterplot: Graphing Correlations

      Positive Correlation

      Negative Correlation
   The Pearson Product-Moment Correlation Coefficient (r)
      Correlation Coefficient
   The Range of r Values
      Perfect Negative Correlation

      Zero Correlation
      Perfect Positive Correlation

II. Review Summary

III. Check Your Progress

IV. Interpreting Correlation Coefficients
      Statistically Significant

Practice Exam

Multiple Choice

Identify the letter of the choice that best completes the statement or answers the question.

_____ 1.   The extent that two variables are related is know as

    a. variance.   c. a correlation.
    b. power.   d. an extraneous variable.

_____ 2.   A _____ is used to graphically display a correlation.

    a. pie chart   c. frequency polygon
    b. histogram   d. scatterplot

_____ 3.   When graphically displaying the relationship between two variables, the
_____ is displayed on the vertical (or Y) axis and the _____ is
displayed on the horizontal (or X) axis.

a. ordinate, abscissa   c. ordinate, intercept
b. abscissa, ordinate   d. slope, abscissa

_____ 4.   An inverse relationship between two variables describes a _____
correlation.

a. positive   c. zero
b. negative   d. neutral

_____ 5.   A single number that expresses the degree of relationship between two variables
is called the

a. variance.   c. standard deviation.
b. correlation coefficient.   d. mean.

_____ 6.   Sherlock computed a correlation coefficient and found that low scores on variable
A were associated with high scores on Variable B. What type of correlation is
this?

a. positive   c. zero
b. negative   d. neutral

_____ 7.   A strong positive correlation is graphically represented by a line that

a. is vertical.   c. slopes downward to the right.
b. is horizontal.   d. slopes upward to the right.

_____ 8.   The numerical range of a correlation coefficient is

a. 0.00 to +1.00.   c. -1.00 to +1.00.
b. -1.00 to 0.00.   d. -10.0 to +10.0.

_____ 9.   Which of the following correlation coefficients demonstrates the strongest
possible relationship?

a. +.60   c. -.70
b. +1.70   d. -.42

88

____ 10. A _____ correlation provides a measure of relation between a nominal variable and an interval variable.

 a. point biserial
 b. phi coefficient
 c. contingency coefficient
 d. Pearson product moment

____ 11. A statistically significant result indicates that a research result occurred

 a. frequently by chance.
 b. rarely by chance.
 c. half of the time.
 d. none of the above

____ 12. Sir Ronald Fisher was responsible for setting the standard for statistical significance at the _____ level.

 a. .25
 b. .10
 c. .05
 d. .01

____ 13. The third step in the process of concluding if your correlation coefficient is statistically significant requires determining

 a. the critical value.
 b. the obtained value.
 c. if the test is one or two-tailed.
 d. the correct degrees of freedom.

____ 14. Which concept describes the size or magnitude of the effect of an IV on a DV?

 a. coefficient of determination
 b. significance level
 c. correlation coefficient
 d. effect size

____ 15. Javier and Samuel are studying two different instructional strategies. What type of information should they use to determine the effectiveness of their instructional strategies?

 a. coefficient of determination
 b. effect size
 c. significance level
 d. z-score distribution

____ 16. Researchers refer to the squared correlation coefficient as a(n)

 a. coefficient of determination.
 b. effect size.
 c. significance level.
 d. regression equation.

____ 17. _____ r^2 values indicate that factors, other than the two variables of interest, are influencing the relation researchers are seeking.

 a. Higher
 b. Lower
 c. Moderate
 d. Negative

18. Which of the following correlation coefficients will account for the most variance?

   a. .102
   b. .257

   c. .758
   d. .828

19. Which correlation coefficient will yield the largest coefficient of determination?

   a. $r = .40$
   b. $r = -.50$

   c. $r = .60$
   d. $r = -.70$

20. A graphical display of the relation between values on the predictor variable and predicted values on the criterion variable is called the

   a. Y intercept.
   b. slope.

   c. regression line.
   d. inflection point.

21. When using regression researchers use the X or _____ variable to predict the Y or _____ variable.

   a. criterion, predictor
   b. extraneous, predictor

   c. extraneous, criterion
   d. predictor, criterion

22. The steepness of the regression line is called the _____; whereas the point at which the regression line crosses the ordinate is called the _____.

   a. slope, intercept
   b. effect size, intercept

   c. intercept, slope
   d. regression, slope

Matching

a. r^2
b. intercept
c. $r^2 = .130$
d. $r = -.07$
e. Y
f. $r = 0.00$
g. regression

h. $r = -.87$
i. $r^2 = .837$
j. X
k. Pearson product moment correlation
l. df
m. slope
n. correlation coefficient

_____ 1. weak negative correlation

__e__ 2. symbol for a criterion variable

_____ 3. a value representing a degree of relation between two variables

__l__ 4. abbreviation for degrees of freedom

__f__ 5. a coefficient indicating a zero correlation

__b__ 6. point where the regression line crosses the Y axis

_____ 7. a coefficient of determination accounting for a large percentage of variance

_____ 8. a coefficient indicating a strong negative correlation

__n__ 9. statistical method for predicting one variable from another

_____ 10. a coefficient of determination accounting for a small percentage of variance

__m__ 11. a term describing the steepness of a regression line

_____ 12. the most common measure of a correlation coefficient

__j__ 13. symbol for a predictor variable

__a__ 14. symbol for the coefficient of determination

True/False

Indicate whether the sentence or statement is true or false.

_____ 1. Correlations indicate the extent to which two variables are related.

_____ 2. Correlations imply causation.

_____ 3. A strong correlation between two variables indicates a cause-and-effect relationship.

_____ 4. The Pearson product-moment correlation coefficient is the most common measure of a correlation.

_____ 5. A zero correlation indicates that there is little or no relation between two variables.

_____ 6. A correlation coefficient of +0.10 indicates that there is a perfect positive correlation.

_____ 7. Sir Ronald Fisher was responsible for setting the common statistical significance criterion at the .01 level.

_____ 8. A statistically significant result provides little information about the magnitude or importance of the result of an experiment.

_____ 9. The regression line is a graphical display of the relation between values of the predictor variable and predicted values on the criterion variable.

_____ 10. The point at which the regression line crosses the Y axis is called the slope.

Short Answer

1. Define the term correlation coefficient. Discuss the relation between negative, positive, and zero correlations. Provide an example drawing of each.

2. Summarize why a correlation coefficient does not imply causation.

3. List and describe the key components needed to construct a correlation scatterplot.

4. Examine and label the components of the raw-score formula for calculating a correlation coefficient (r). List the necessary steps involved with this calculation.

5. Explain what researchers mean by the statement "rarely by chance." Distinguish between a .01 and .05 level of significance.

6. List the steps used to determine if a correlation coefficient is statistically significant.

7. Summarize why it is important for researchers to understand and interpret the effect size of their result.

8. Discuss the purpose of the coefficient of determination and describe how it is interpreted.

9. Label and describe the components of the regression equation.

10. Define and distinguish between the slope and the Y intercept in the regression equation.

SPSS Computer Practice Problems 8.1 and 8.2 Using SPSS 11.0 for Windows

Basically when conducting a statistical analysis using SPSS for Windows you will name your variables, enter your data, and finally analyze your data by selecting options from a toolbar. In the following section (i.e., problem 8.1) a step by step procedure of an example problem will be provided to guide you in calculating a correlation and constructing a corresponding scatterplot to graphically display your data. By following the systematic instructions provided and by referring to the screen figures (SF) when prompted, you should be able to independently conduct a correlation analysis and construct a corresponding scatterplot. Practice problem 8.2 will follow problem 8.1 and will help you to conduct a linear regression analysis.

Example Computer Problem 8.1

Data

The fictional data for this problem is based on an economist who studied the relation between years of education and income levels. The data is based on responses from 15 participants. The variables include the participant's educational level or years of education (variable X) and the participant's current income (variable Y). The raw data for the participants is provided in Table 1.

Table 1

Participants	Years of Education (X)	Income Level (Y)
1	11	$15,900
2	16	$18,000
3	12	$14,500
4	15	$13,000
5	6	$11,600
6	9	$12,700
7	17	$17,700
8	16	$15,900
9	20	$20,400
10	14	$14,000
11	10	$13,500
12	13	$17,100
13	8	$14,200
14	12	$16,300
15	14	$16,800

While working on example problem 8.1 consider the following questions:

- What is the Pearson product-moment correlation coefficient? Is there a strong, moderate, or weak relationship between the variables?

- Is the Pearson product-moment correlation coefficient significant? If so, at what alpha level (i.e., .05 or .01) is it significant?

- Is this a positive, negative, or zero order correlation?

- What is the coefficient of determination for this analysis? What does the coefficient of determination tell you about the relationship of the data?

- What is the direction and shape of the scatterplot? Are their any outliers present in the data set?

Procedure for Completing Example Problem 8.1

1. **Getting Started.** Open the SPSS 11.0 program
2. **Data Option.** At this point the SPSS window should provide a prompt asking, "What would you like to do?" Using your mouse click on the radial button to "Type in data."
3. **Data Entry.** Next you will begin to enter the data. This requires manually typing each raw score into a column (see Screen Figure 8.1 or SF-8.1). For example, you will enter the first participant's education level in the first data cell of row 1 for column 1 (i.e., 11.00) and then you will enter this participant's corresponding income level in the first data cell of row 1 for column two (i.e., 15,900). At this point our variables are named "var00001" (education level) and "var00002" (income level), however we will change these shortly. Next you will enter the second participant's education level and corresponding income (i.e., 16 and 18,000) and so on until you have entered all of the data for the 15 participants (see SF-8.1). It does not matter which variable you place in column one or column two just as long as the corresponding data for each participant is placed side by side in each row.

SF-8.1. Entering data to conduct a correlation analysis.

4.   **Naming Your Variable(s).** At this point it may be helpful to rename or replace the default variables ("var00001" and "var00002") by selecting or clicking on the variable view tab at the bottom left hand corner of the screen (see SF-8.1). Naming your variables will be especially important for organization and record keeping when you begin to add more variables to your data set. After selecting the variable view tab you should be able to view the variables named "var00001" and "var00002." Under the column "Name" in the first cell you can provide an alternative name for your variables by highlighting or selecting the cell with your mouse. At this point your variable cell should be highlighted with a black box. You may use up to eight characters to name your variable(s). Each row on the left hand side of the screen, indicated by sequential numbering, represents a potential variable. For this example I have decided to name my variables "educatio" and "income" to more accurately explain my data (see SF-8.2). Once you are finished select the "Data View" tab at the lower left hand corner of the screen and this will prompt SPSS to return you to the data entry screen.

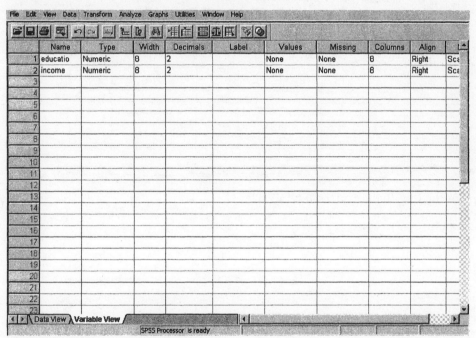

SF-8.2. Naming your variables.

5. **Conducting the Correlation Analysis.** In order to conduct a statistical analysis using SPSS you will need to select the "Analyze" toolbar option at the top of the screen. A dropdown box will appear providing further data analysis options. Select the "Correlation" option. After selecting this option SPSS will provide you with further analysis alternatives to the right of the initial dropdown box (see SF-8.3). Next select the "Bivariate" option.

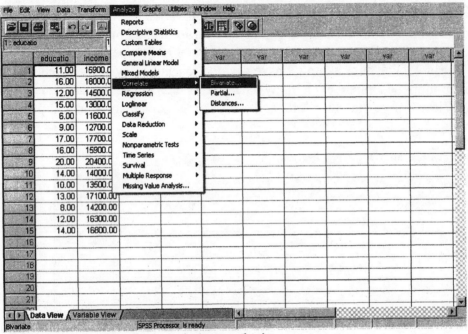

SF-8.3. Conducting the correlation analysis.

After selecting the "Bivariate" option you should be able to view a smaller command screen labeled "Bivariate Correlations" (see SF-8.4). Select or highlight the variables (in this case "educatio" and "income") by clicking on the variable names and then use the arrow button located in the middle of the screen to move your variables into the variables box. Once you have selected your variables you can select the type of correlation analysis you want to conduct. For the purpose of this example we will use the "Pearson" or Pearson Product-Moment correlation. This option can be selected by clicking on the open box to the left of the "Pearson" category under the "Correlation Coefficients" heading. A checkmark will indicate that you have properly selected this option (see SF-8.4). Next you will need to select the "Flag significant correlations" option, located in the bottom left corner, by clicking on the open box. Lastly, in order to begin the correlation analysis you will need to select the "OK" button located on the top right corner of the "Bivariate Correlations" command box. Once selected, SPSS will conduct the correlation analysis.

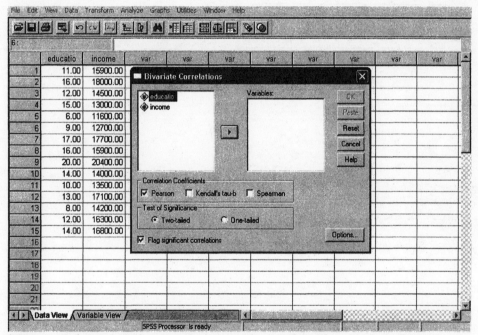

SF-8.4. Conducting the correlation analysis: Selecting your variables.

6. **Reviewing the Output for the Correlation Analysis.** Your output should resemble the output in SF-8.5. If it does not you may want to check your raw data for data entry errors. The output in Screen Figure 8.5 provides a correlation matrix for your variables. It also provides information on the number of participants in the analysis ($N = 15$), the correlation coefficient ($r = .786$), and indicates that the result was significant at the .01 level.

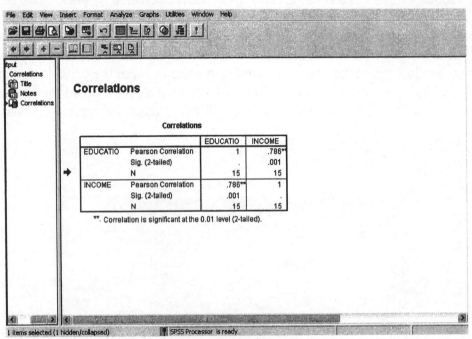

SF-8.5. Output for the correlation analysis.

7. **Creating the Scatterplot.** Next you will need to graphically display your statistical output by creating a scatterplot of the data. This can be accomplished by selecting the "Graphs" option located at the top of the SPSS command screen (see SF-8.6). Once selected a dropdown command box will appear. At this point you should select the "Scatter..." option. Now you should be able to view a scatterplot option box (see SF-8.7).

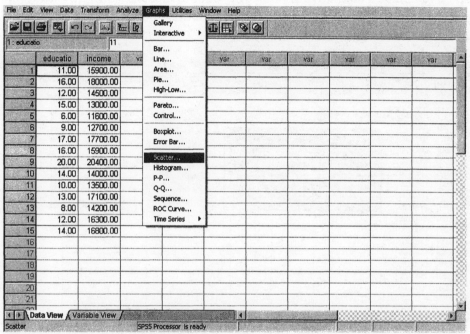

SF-8.6. Creating the scatterplot.

Next you will need to select the type of scatterplot you want to graphically display your data. For the purpose of this example use the "Simple" scatterplot option. This option can be selected by clicking on the "Simple" scatterplot picture (see SF-8.7). After selecting this option you will notice that the "Simple" scatterplot option is highlighted in black. Then click on the "Define" button, located in the right hand corner of the "Scatterplot" command box, to verify that you want to create a "Simple" scatterplot. After selecting the "Define" command another command box will appear with the heading "Simple Scatterplot" (see SF-8.8).

SF-8.7. Creating the scatterplot: Selecting options.

At this point you should be able to view the "Simple Scatterplot" command screen (see SF-8.8). This command screen will allow you to select which variables you want placed on the X and Y axes. To begin, select a variable by clicking on the variable. Once the variable is highlighted in blue you can move it to the X or Y axis box by selecting the arrow button to the left of the X or Y axis boxes. For the purpose of this example place the variable "income" on the Y axis and the variable "educatio" on the X axis. Next select the "OK" command button located in the upper right hand corner of the screen. This will prompt SPSS to create your scatterplot.

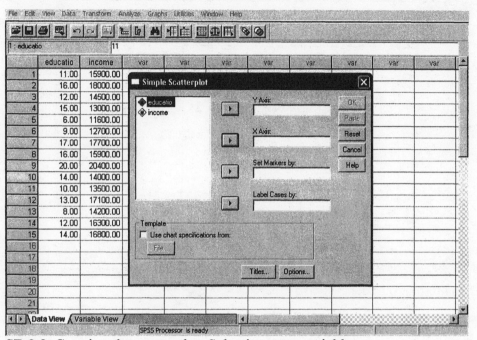

SF-8.8. Creating the scatterplot: Selecting your variables.

8.   **Reviewing the Scatterplot Output**. The results of your scatterplot should resemble the output in SF-8.9. The output graphically displays a data point for each participant. Notice that there are 15 data points plotted on the scatterplot. Each data point represents information about the participant's level of education and their corresponding income level.

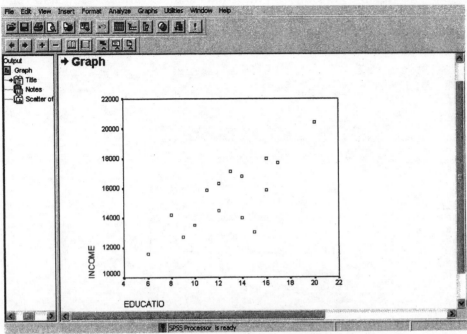

SF-8.9. Scatterplot output.

9. Interpreting the Results.

Use Table 2 and Figure 1 to help you answer the questions below.

Table 2

Correlation Output

Correlations

		EDUCATIO	INCOME
EDUCATIO	Pearson Correlation	1	.786**
	Sig. (2-tailed)	.	.001
	N	15	15
INCOME	Pearson Correlation	.786**	1
	Sig. (2-tailed)	.001	.
	N	15	15

**. Correlation is significant at the 0.01 level (2-tailed).

Figure 1

Scatterplot

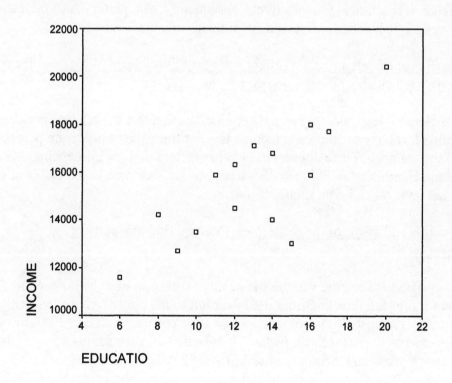

What was the Pearson Product-Moment correlation coefficient? Was there a strong, moderate, or weak relationship between the variables?

The Pearson Product-Moment correlation coefficient was $r = .786$. According to Best and Khan (1989), the strength of the relationship between the variables was substantial. Best and Khan utilize the following criterion to interpret the strength of a correlation:

Negligible   ($r = .00$ to $.20$)
Low   ($r = .20$ to $.40$)
Moderate   ($r = .40$ to $.60$)
Substantial   ($r = .60$ to $.80$)
High to very high   ($r = .80$ to 1.00)

Best, J.W., & Kahn, J. V. (1989). *Research in Education* (6th ed.) Englewood Cliffs, NJ: Prentice Hall.

Was the Pearson Product-Moment correlation coefficient significant? If so, at what alpha level (i.e., .05 or .01) was it significant?

The correlation coefficient ($r = .786$) was significant at the .01 level.

Was this a positive, negative, or zero order correlation?

The correlation coefficient was positive. Consequently, as a participant's educational level (or years of education) increased, so did their income level.

Calculate the coefficient of determination (effect size) for this analysis. What does the coefficient of determination (r^2) tell us about the relationship of the data?

The coefficient of determination was $r^2 = .617$ indicating that the relation between a participant's level of education and income level accounts for a moderate percentage (61.7) of the variance. There appears to be other factors that are influencing this relationship. Hence, the ability to predict a participant's income based on his or her educational level for this data should be fair.

Describe the direction and shape of the scatterplot. Can you find any outliers present in this data set?

The relationship between the variables is positive and somewhat linear because the participant scores fall on a line from the lower left to the upper right of the scatterplot (see Figure 1). Because this is not a perfect positive correlation (i.e., +1.00) the scores do not follow an exact upward linear pattern. It appears that participants #5 (education = 6 years and income = $11,600) and #9 (education = 20 years and income = $20,400) may be outliers given that they are somewhat removed from the general pattern of scores.

Example Computer Problem 8.2

Data

The data for this fictional problem is based on a college administrator who wants to predict freshman grade point averages (GPA) based on their college entrance exam scores (ACT) during the second semester of school. The data is based on GPA and ACT scores for 15 freshmen. The criterion variable ($\hat{y}$) or the variable that is being predicted in this example is student GPA scores. The predictor variable (x) or the variable we are predicting from is student ACT scores. The raw data for the students is provided in Table 1.

Table 1

Participants	ACT exam scores (X)	Student GPA ($\hat{y}$)
1	32	3.29
2	23	2.17
3	17	1.58
4	32	4.00
5	24	2.53
6	28	2.89
7	20	1.92
8	31	3.85
9	22	2.05
10	21	1.70
11	34	3.70
12	26	2.77
13	30	3.51
14	27	2.26
15	29	3.15

Procedures for Completing Example Problem 8.2

Please refer to the previous problem (8.1) for instructions on getting started, entering your data, and naming your variables.

1. **Conducting the Linear Regression Analysis.** In order to conduct a statistical analysis using SPSS you will need to select the "Analyze" toolbar option at the top of the screen. A dropdown box will appear providing further data analysis options. Select the "Regression" option. After selecting this option SPSS will provide you with further analysis alternatives to the right of the initial dropdown box. Choose the command "Linear" for your analysis (see SF-8.1).

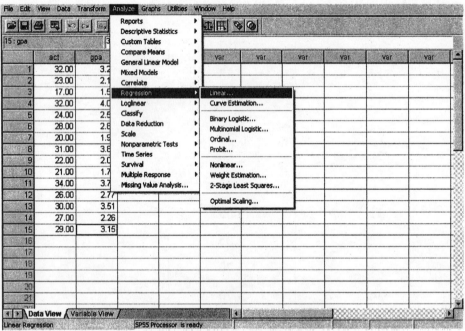

SF-8.1. Conducting the linear regression analysis.

At this point you should be able to view a smaller command screen labeled "Linear Regression" (see SF-8.2). Select or highlight the variables (in this case "act" and "gpa") by clicking on the variable names and then use the arrow buttons located in the middle of the screen to move your variables into the appropriate variable box (i.e., "Dependent" or "Independent(s)"). For the purpose of this example "gpa" will be our DV and "act" will be our IV, given that we want to predict a student's GPA based on their ACT exam score (see SF-8.2). Once you have selected and assigned your variables you can select the "OK" button located on the top right section of the "Linear Regression" command box. Once selected, SPSS will conduct the linear regression analysis.

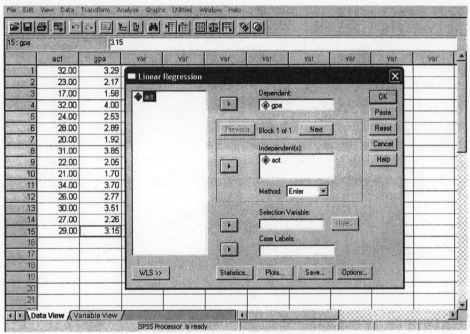

SF-8.2. Conducting the linear regression analysis: Selecting your variables.

2. **Reviewing the Output for the Linear Regression Analysis.** Your output should resemble the output in SF-8.3. If it does not you may want to check your raw data for data entry errors. The output in Screen Figure 8.3 provides information about **R, R Square, Adjusted R Square,** and the **Std. Error of the Estimate**. Although there is a great deal of detailed information that is used for advanced analyses we will only pay attention to the values for **R** and **R Square**. When using SPSS for bivariate prediction, the value for **R** is actually the ordinary bivariate correlation coefficient (*r*) and the **R Square** value is actually the r^2 (squared correlation coefficient) or coefficient of determination. In this example **R** = .942 and **R Square** = .887.

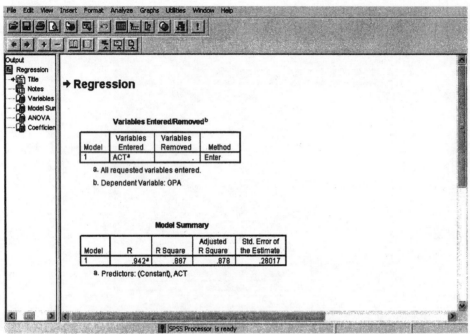

SF-8.3. Output for the linear regression analysis: Model summary.

If you scroll down to the next screen, your results should resemble the output displayed in Screen Figure-8.4. The information needed to construct our raw score prediction model can be found in the "Coefficients" output box. The first column on the left side of this box titled "Model" displays our predictor variable (ACT). In the second column, first box, titled "B" SPSS provides the information we need to construct our raw score prediction model. The first value (**-1.174**) provided is the raw score regression constant or the "*a*" in our regression equation ($\hat{y} = a + bx$). The second value (**.149**) provided is the raw score regression coefficient or the *b* in our regression equation. If we plug this information into our regression equation we will have our raw score prediction model as indicated below:

GPA ($\hat{y}$) = -1.174 (*a*) + .149 (*b*) ACT (*x*)

In the third major column titled "Standardized Coefficients" SPSS provides the standardized regression coefficient or **Beta**. In this example, Beta is equal to **.942**. In the last column titled "Sig" SPSS provides information about statistical significance. In general, if the value in the predictor column (ACT) is less than .05 the result is considered to be statistically significant. In this example our result is significant because *p* < .01.

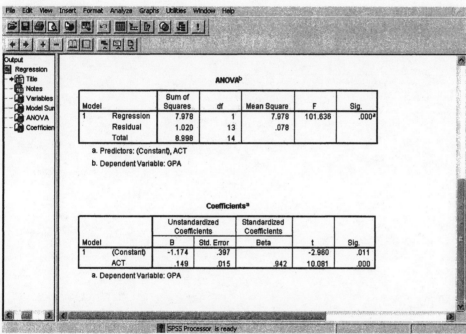

SF-8.4. Output for the linear regression analysis: Coefficients.

3. Applying the Results.

$$\text{GPA } (\hat{y}) = -1.174 \ (a) + .149 \ (b) \ \text{ACT } (X)$$

Now that we have our raw score prediction model try answering the following questions:

1. If a student has an ACT score of 24, what would you predict their GPA to be?

2. If a student had an ACT score of 15, what would you predict their GPA to be?

3. If a student had an ACT score of 31, what would you predict their GPA to be?

Answers to the multiple choice, matching, true/false items, and practice problems

Multiple Choice		Matching		True and False		Practice Problem 8.2
1.	C	1.	D	1.	T	1. GPA = 2.40
2.	D	2.	E	2.	F	2. GPA = 1.06
3.	A	3.	N	3.	F	3. GPA = 3.44
4.	B	4.	L	4.	T	
5.	B	5.	F	5.	T	
6.	B	6.	B	6.	F	
7.	D	7.	I	7.	F	
8.	C	8.	H	8.	T	
9.	C	9.	G	9.	T	
10.	A	10.	C	10.	F	
11.	B	11.	M			
12.	C	12.	K			
13.	A	13.	J			
14.	D	14.	A			
15.	B					
16.	A					
17.	B					
18.	D					
19.	D					
20.	C					
21.	D					
22.	A					

Chapter 9

Probability, the Normal Curve, and Sampling

Learning Objectives

By the end of this chapter you should be able to:

1. Summarize the nature and logic of using probability to make decisions about your inferential statistics.

2. Compare and contrast the purposes for and applications of z and t-tests.

3. Explain the purpose and shape of a t-distribution.

4. Formulate a hypothesis test for a one-sample t-test analysis.

5. Calculate and interpret a t-obtained value for a one-sample t-test.

6. Summarize the statistical results of a one sample t-test using the format of the American Psychological Association's Publication Manual.

7. Distinguish between a one-tailed and two-tailed test of significance.

8. Define and describe the problems associated with Type I and Type II errors.

9. Compare and contrast sampling considerations and basic research strategies.

By the end of this chapter you should learn when and how to use the following formula:

1. <u>Formula for calculating a one-sample t-test.</u>

$$t_{obt} = \frac{M - \mu}{SEM} \quad \text{and} \quad SEM = \frac{SD}{\sqrt{N}}$$

Where:

M	=	sample mean
μ	=	population mean
SEM	=	standard error
SD	=	standard deviation
N	=	number of participants

Expanded Outline

I.   The Nature and Logic of Probability
                Probability

                Subjective Probability

                Percentages
                Proportions
                Null Hypothesis

                Alternative Hypothesis

           A Conceptual Statistical Note
           Probability and the Normal Curve
                Normal Distribution
                Standard Deviation
                z-score
                Parameter
           Probability and Decisions

           Comparing a Sample to a Population: The One-sample *t*-Test
                t-distribution
                Critical Region

                Degrees of Freedom

                Power

                Marginal Significance
           One-tailed Versus Two-tailed Tests of Significance
                Two-tailed Test
                One-tailed Test

           When Statistics Go Astray: Type I and Type II Errors
                Type I Error

                Type II Error

II.   Review Summary

115

Practice Exam

Multiple Choice

Identify the letter of the choice that best completes the statement or answers the question.

_____ 1. Which of the following numerical expressions can be used to describe probability?

    a. proportions   d. all of the above
    b. odds   e. none of the above
    c. percentages

_____ 2. The _____ hypothesis assumes that all differences between participant groups are due to chance and not the operation of the IV; whereas the _____ hypothesis assumes that all differences between the participant groups are not due to chance, but to the effect of the IV.

    a. alternative; null   c. primary; null
    b. null; alternative   d. null; primary

_____ 3. A null hypothesis can be mathematically expressed as:

    a. $M_1 - M_2$ c. $M_1 = M_2$
    b. $M_1 \neq M_2$ d. $M_1 + M_2$

_____ 4. Inferential statistics with small values occur _____ by chance; whereas inferential statistics with large values occur _____ by chance.

    a. frequently; rarely   c. equally; sporadically
    b. rarely; frequently   d. none of the above

_____ 5. The proportion of times an event would occur if the chances for occurrence were infinite best describes

    a. the addition rule of probability.   c. probability.
    b. conditional probability.   d. a gambler's fallacy.

_____ 6. Traditionally researchers have accepted that any event that occurs by chance _____ times or fewer in 100 occasions is a rare event.

    a. 5   c. 12
    b. 10   d. 15

_____ 7. A probability based on an individual's unique perspective is called a(n)

a. theoretical probability.
b. empirical probability.
c. real-world probability.
d. subjective probability.

_____ 8. In comparison to a normal distribution, the _t_-distribution uses samples that are typically

a. larger.
b. smaller.
c. random.
d. non-random.

_____ 9. The shape of a _t_-distribution

a. is not typically normal.
b. is not symmetrical.
c. changes with the size of the distribution.
d. is negatively skewed.

_____ 10. The ability of a number in a given set to assume any value is called

a. degrees of freedom.
b. power.
c. the critical region.
d. marginal significance.

_____ 11. A researcher wants to compare a sample mean to a population mean. What inferential statistic should be used?

a. a one-sample _t_-test
b. a _z_-score
c. an analysis of variance
d. a correlation coefficient

_____ 12. Professor Parker examined the difference between the formula for a one-sample _t_-test and a _z_-score formula. He found that the one-sample _t_-test formula used a _____ to calculate the _t_-score; whereas the _z_-score formula used a _____ to calculate a _z_-score.

a. single raw score and SD; sample mean and SEM
b. sample mean and SD; single raw score and SEM
c. sample mean and SEM; single raw score and SD
d. single raw score and SEM; sample mean and SD

_____ 13. Before using a _t_-test researchers should assume that the

a. population is skewed.
b. population is normally distributed.
c. sample is normally distributed.
d. sample is skewed.

_____ 14. A research article in the *Psi-Chi Journal of Undergraduate Research* reports the results of a study using a t-test as: $t(50) = 2.69, p < .05$. This finding indicates that the

a. result is not significant at the .05 level. c. the t-score was 2.69.
b. the t-score was 50.   d. there were exactly 50 participants.

_____ 15. Archibald has conducted a statistical analysis of his data using a t-test. What should he conclude if the result of his analysis yielded the following output: $t(10) = 3.35, p < .05$.

a. reject the null hypothesis   c. it is not significant at the .05 level
b. fail to reject the null hypothesis   d. the result is inconclusive

_____ 16. You find that your obtained t-value is 3.24. Then you look up the corresponding critical t-value for this result using a t-table and find that the critical t-value is 3.745. What should you conclude?

a. your result is significant   c. your result was inconclusive
b. your result is not significant   d. none of the above

_____ 17. Statistical findings that have a probability of chance between .05 and .10 are referred to as

a. fortunate.   c. descriptively accurate.
b. stringent when compared to a .01 level.   d. marginally significant.

_____ 18. Directional hypotheses are associated with _____ t-tests and non-directional hypotheses are associated with _____ t-tests.

a. one-tail; two-tail   c. two-tail; three-tail
b. two-tail; one-tail   d. one-tail; three-tail

_____ 19. Gertrude predicts that student scores in her class will be higher than the national average. How should she express her experimental hypothesis?

a. $M_1 = M_2$ c. $M_1 > M_2$
b. $M_1 \neq M_2$ d. $M_1 - M_2$

_____ 20. Researchers should provide a strong rational for predicting the direction of their experimental or alternative hypothesis before using a

a. one-tailed test.   c. t-test for independent samples.
b. two-tailed test.   d. t-test for dependent sample.

21. Which of the following is used as a level of rejection by researchers when testing a hypothesis?

   a. descriptive level
   b. beta level

   c. alpha level
   d. *t*-score level

22. Accepting the alternative hypothesis when the null hypothesis is true is referred to as a

   a. Type I error.
   b. Type II error.

   c. Type III error.
   d. none of the above

23. Accepting the null hypothesis when the alternative hypothesis is true is referred to as a

   a. Type I error.
   b. Type II error.

   c. Type III error.
   d. none of the above

24. Typically researchers set the alpha level at _____ because it places the probability of type I and type II errors at acceptable levels.

   a. .01
   b. .05

   c. .10
   d. .15

25. A researcher hypothesizes that the mean income for a population is $32,500. The population mean income is actually $32,500 and the researcher rejects the hypothesis. What type of error has the researcher made?

   a. Type I
   b. Type II

   c. Type III
   d. the researcher did not make an error

26. Leopold has taken great care so that every member of the population has an equal likelihood of being selected for inclusion into his sample. What sampling procedure is he using given that he does not want the same participants to appear more than once in a participant group?

   a. non-random sampling
   b. stratified random sampling

   c. random sampling with replacement
   d. random sampling without replacement

27. The sampling procedure that divides a population into subpopulations and then draws a random sample from one or more of these subpopulations is referred to as

   a. un-stratified random sampling.
   b. stratified random sampling.

   c. random sampling with replacement.
   d. random sampling without replacement.

_____ 28. Which research strategy seeks to answer rather specific questions by acquiring data from a single, specified segment of the population?

    a. modified sampling   c. the single-strata approach
    b. longitudinal research   d. cross-sectional research

Matching

a. stratified random sampling   h. single-strata approach
b. normal distribution   i. β
c. $M_1 \neq M_2$ j. random sampling with replacement
d. parameter   k. critical region
e. α l. random sampling without replacement
f. degrees of freedom   m. $M_1 = M_2$
g. $M_1 > M_2$ n. _t_-distribution

_____ 1. random samples drawn from specific subpopulations

_____ 2. a statistical value for a population

_____ 3. resembles a bell-shaped distribution

_____ 4. a null hypothesis expression for two participant groups

_____ 5. gathering data from a single stratum of a selected population

_____ 6. the symbol for a Type II error

_____ 7. an expression for a directional hypothesis

_____ 8. the ability of a number in a given set to assume any value

_____ 9. a participant cannot be returned to the population once randomly selected

_____ 10. a symmetrical distribution with half of the scores above and below the mean

_____ 11. usually the extreme area (.05) of a distribution

_____ 12. the symbol for a Type I error

_____ 13. a participant can be returned to the population once randomly selected

_____ 14. an experimental hypothesis expression for two participant groups

True/False

Indicate whether the sentence or statement is true or false.

_____ 1. Probability is fundamental to descriptive statistics.

_____ 2. The null hypothesis assumes there is no difference between the participant groups.

_____ 3. As a precaution against bias, researchers assume the alternative hypothesis and let the data and statistical test demonstrate otherwise.

_____ 4. The results of an inferential statistical test can tell us whether the results of an experiment would occur frequently or rarely by chance.

_____ 5. A statistical value for an entire population is called a sample.

_____ 6. The *t*-distribution is typically taller in the middle and flatter on the tails when compared to the normal distribution.

_____ 7. The *t*-distribution changes shape somewhat based on the size of the sample.

_____ 8. Researchers typically report their results at the lowest possible level of chance.

_____ 9. A Type I error involves rejecting a true experimental hypothesis.

_____ 10. The experimenter directly controls the probability of making a Type I error by setting the significance level.

_____ 11. Researchers set the alpha level at .10 because it places the probability of Type I and Type II errors at acceptable levels.

_____ 12. Researchers can indirectly reduce Type II errors by implementing techniques that will cause the participant groups to differ as much as possible.

_____ 13. A one-tailed test is a directional test of significance.

Short Answer

1.   Summarize the rationale behind the assumptions of the null and alternative hypotheses used in research. How are the null and alternative hypotheses mathematically expressed when using two participant groups? Explain what outcome researchers typically seek when using the null and alternative hypotheses.

2.   Compare and contrast a parameter and a sample. Is it possible for a parameter to be the same as a sample? Explain your answer.

3.   Define the terms z-score and t-score. In what situations are they used? In general, what would you expect the shape of a distribution based on a z-score and a t-score to look like?

4.   Explain why the t-distribution changes shape accordingly with the size of your sample.

5.   Define the term degrees of freedom. What is the relationship between the critical values in a t-table and the number of degrees of freedom?

6.   Describe the differences between a one-sample t-test formula and a z-score formula.

7.   Explain the relationship between an obtained t-value and a critical t-value when determining if your result is significant.

8.   Distinguish between a one-tailed and a two-tailed test of significance and describe how they are related to directional and non-directional hypotheses. Try to mathematically express a null and an alternative hypothesis for a one-tailed and a two-tailed test of significance.

9.   Summarize the roles that Type I and Type II errors play in hypothesis testing. Can researchers avoid making Type I and Type II errors? Explain your answer.

10.   Describe how Type I errors, Type II errors, and power affect the decisions and interpretations that researchers make.

11.   Discuss the benefits and limitations of using the stratified random sampling technique. Explain what may happen if your sample becomes too highly specified or stratified.

Answers for the multiple choice, matching, and true/false items

Multiple Choice		Matching		True and False	
1.	D	1.	A	1.	F
2.	B	2.	D	2.	T
3.	C	3.	N	3.	F
4.	A	4.	M	4.	T
5.	C	5.	H	5.	F
6.	A	6.	I	6.	F
7.	D	7.	G	7.	T
8.	B	8.	F	8.	T
9.	C	9.	L	9.	F
10.	A	10.	B	10.	T
11.	A	11.	K	11.	F
12.	C	12.	E	12.	T
13.	B	13.	J	13.	T
14.	C	14.	C		
15.	A				
16.	B				
17.	D				
18.	A				
19.	C				
20.	A				
21.	C				
22.	A				
23.	B				
24.	B				
25.	A				
26.	D				
27.	B				
28.	C				

Chapter 10

Designing and Conducting Experiments with Two Groups

Learning Objectives

By the end of this chapter you should be able to:

1. Summarize the basic components for designing a research study.

2. Explain the role of the Principle of Parsimony in research.

3. Compare and contrast the benefits and limitations of the most basic experimental designs.

4. Define and distinguish between random assignment and random selection.

5. Discuss the benefits and limitations of random and non-random assignment of participant groups.

6. List and summarize the three most common methods for using correlated assignment of participants.

7. Discuss several factors that researchers should consider before selecting an independent or correlated two-groups design.

8. Describe factors that influence within-group and between-group variance.

9. Explain the benefits and limitations of conducting ex post facto research.

Expanded Outline

I.   Experimental Design: The Basic Building Blocks

          Two-Group Design

               Principle of Parsimony

               How many IVs?

                    IVs

                    DVs

               How Many Groups?

                    Extraneous Variables

                    Levels

                    Experimental Group

                    Control Group

           Assigning Participants to Groups

               Random Assignment

               Random Selection

               Independent Groups

               Between-Subjects Comparison

               Confounded Experiment

               Correlated Assignment

                    Matched Pairs

                    Repeated Measures

                    Natural Pairs

                       Within-Subjects Comparison

II.   Review Summary

III.   Check Your Progress

Multiple Choice

Identify the letter of the choice that best completes the statement or answers the question.

_____ 1. When researchers seek to create the simplest possible research design that will yield a valid experiment they are adhering to

   a. ex post facto research.
   b. the Law of Parsimony.
   c. Ockham's razor.
   d. the scientific principle.
   e. both b and c

_____ 2. The simplest experimental design that demonstrates the effect an IV had on the DV is the

   a. two-group design.
   b. multiple correlated group design.
   c. one-way factorial design.
   d. nested repeated measures design.

_____ 3. The first question a researcher should ask when selecting an appropriate design for their experiment is:

   a. How many participant groups?
   b. What type of participant group?
   c. How many DVs?
   d. How many IVs?

_____ 4. In common two-group experimental designs the levels of the IV are often characterized as

   a. strong vs. weak.
   b. high vs. low.
   c. present vs. absent.
   d. parsimonious vs. complex.

_____ 5. The participant group that receives the IV is typically referred to as the _____ group; whereas the participant group that does not receive the IV is referred to as the _____ group.

   a. experimental, control
   b. experimental, nuisance
   c. control, experimental
   d. control, extraneous

_____ 6. Jill has devised an experiment in which one participant group receives a new multivitamin and the other participant group receives a sugar pill or placebo. The participant group receiving the sugar pill is referred to as the _____ group.

   a. experimental
   b. control
   c. nuisance
   d. comparison

____ 7. When selecting participants for a research study, _____ involves choosing your research participants; whereas _____ involves placing those participants into their comparison groups.

a. correlated assignment, random selection
b. random assignment, random selection
c. random selection, correlated assignment
d. random selection, random assignment

____ 8. A researcher tests the effects of a weight loss drug on participants' by administering 1, 2, 3, or 4 tablets of the drug per day. How many IVs were used in this study?

a. 1
b. 2
c. 3
d. 4

____ 9. Sunshine is studying the influence of different therapies for depression. She chooses to investigate the effects individual counseling, group counseling, and medication therapy will have on depression. How many levels does her IV have?

a. 1
b. 2
c. 3
d. 4

____ 10. When researchers randomly assign participants to groups, they have created

a. biased groups.
b. independent groups.
c. dependent groups.
d. natural sets.

____ 11. Which statement describes the basic strategy of randomly assigning participants to groups?

a. independent groups design
b. randomized groups design
c. between-subjects design
d. all of the above

____ 12. Ken has decided to use random assignment hoping to create equal participant comparison groups. However, he is unsure about how many participants he should obtain. How many participants should he select to be more confident that he has obtained equal groups?

a. 25
b. 50
c. 75
d. 100

____ 13. Which term is <u>not</u> related to the correlated assignment of participant groups?

      a. matched
      b. non-random

      c. random
      d. paired assignment

____ 14. Cynthia is using a two-group research design. She has measured and equated her participants using a pre-selected variable before her experiment. Which method of correlated assignment is she likely using?

      a. matched pairs
      b. repeated measures

      c. natural pairs
      d. related measures

____ 15. Research participants in a two-group design who are naturally related in some way are called

      a. matched pairs.
      b. repeated measures.

      c. natural pairs.
      d. related measures.

____ 16. Within-group variability is commonly referred to as

      a. between-groups variability.
      b. error variability.

      c. ex post facto variability.
      d. a true experiment.

____ 17. Which of the following is <u>not</u> considered to be a common method for using correlated assignment?

      a. repeated measures
      b. natural pairs

      c. independent groups
      d. matched pairs

____ 18. A researcher has used a two-correlated groups design when the participants are

      a. matched on a pre-test variable.
      b. observed or measured twice.
      c. biologically or socially related.

      d. both a and b
      e. all of the above

____ 19. As the size of a sample increases, it becomes more likely that _____ assignment will create equal groups.

      a. random
      b. correlated

      c. non-random
      d. selective

____ 20. _____ designs can statistically benefit researchers because they can help reduce error variation.

a. Random assignment   c. Non-random
b. Correlated-groups   d. Selective

____ 21. The goal in a research experiment is to maximize the _____ groups variability and minimize the _____ groups variability.

a. between, within   c. within, between
b. between, correlated   d. within, correlated

____ 22. Error variability is also known as

a. individual differences. d. within-groups variability.
b. measurement error.   e. all of the above
c. extraneous variation.

____ 23. A measured IV is associated with a(n)

a. nuisance variable.   c. true experiment.
b. extraneous variable.   d. ex post facto experiment.

____ 24. The degrees of freedom for the independent-groups design is equal to the total number of participants minus _____; whereas the degrees of freedom for the correlated-groups design is equal to the number of pairs of participants minus _____.

a. 1, 2   c. 2, 2
b. 2, 1   d. 3, 1

____ 25. The primary advantage of an independent-groups design is

a. accuracy.   c. simplicity.
b. validity.   d. reliability.

____ 26. Javier has a small number of participants and he expects the effect of his IV to be minimal. What type of two-group design should he use?

a. dependent-groups   c. random-groups
b. independent-groups   d. correlated-groups

_____ 27. When researchers conduct an experiment contrasting different amounts of an IV they no longer have a true

        a. control group.   c. ex post facto experiment.
        b. experimental group.   d. experimental design.

_____ 28. A researcher who is <u>not</u> able to manipulate an IV is conducting a(n)

        a. true experiment.   c. confounded study.
        b. ex post facto study.   d. both b and c

Matching

a. experimental design   h. independent groups
b. random assignment   i. true experiment
c. error variability   j. repeated measures
d. control group   k. levels
e. ex post facto research   l. random selection
f. between-groups variability   m. experimental group
g. degrees of freedom

_____ 1. ensures that each participant has an equal chance of being in any group

_____ 2. the ability of a number in a specified set to assume any value

_____ 3. variability in DV scores due to factors other than the IV

_____ 4. participants who receive the IV in a two-group design

_____ 5. participant group formed by random assignment

_____ 6. variability in DV scores due to the effects of the IV

_____ 7. participants are tested or measured more than once

_____ 8. also known as treatment conditions

_____ 9. participants who do <u>not</u> receive the IV in a two-group design

_____ 10. an experiment in which the researcher directly manipulates the IV

_____ 11. an experiment in which the researcher cannot directly manipulate the IV

_____ 12. ensures each member of the population has an equal chance of being selected

_____ 13. also known as a research blueprint

True/False

Indicate whether the sentence or statement is true or false.

____ 1. Believing that the explanations for events should remain simple until they are no longer valid is called the Principle of Parsimony.

____ 2. Levels are used to describe the differing amounts or types of an IV used in an experiment.

____ 3. The general plan for selecting and assigning participants to experimental conditions, controlling extraneous variables, and gathering data is known as ex post facto research.

____ 4. In a two-group design, the participant group that receives the manipulated IV is known as the control group.

____ 5. Within-groups variability is also known as error variability.

____ 6. Correlated assignment is a method for assigning participants to groups so that there is a relationship between a small number of participants.

____ 7. Independent participant groups are formed using non-random methods.

____ 8. In a two-group design, the participant group that <u>does</u> <u>not</u> receive the manipulated IV is known as the experimental group.

____ 9. The two-group design is the most parsimonious design available to researchers.

____ 10. Natural pairs refers to research participants in a two-group design who are measured and equated on some variable before the experiment.

____ 11. A confounded experiment makes drawing cause-and-effect relationships easier.

____ 12. Although randomization is designed to equate participant groups, the methods for creating correlated-groups designs provide researchers with greater certainty of equality.

____ 13. Smaller inferential statistical values are more likely to demonstrate significant differences.

Short Answer

1.   Explain the role of the Principle of Parsimony in research.

2.   Describe the most basic experimental design. What are the benefits and limitations of this design?

3.   List and describe the three primary questions a researcher should consider, regarding experimental designs, before conducting an experiment.

4.   Define and discuss the differences between the terms random assignment and random selection.

5.   Summarize the benefits and limitations of randomly assigning participant groups.

6.   Compare and contrast the benefits and limitations of non-random assignment of participant groups.

7.   List and describe the three common methods for using correlated assignment of participant groups. Are these methods related to correlation coefficients? Explain your answer.

8.   Describe a situation in which the method of matched assignment is guaranteed to create equal participant groups when the method of random assignment is not.

9.   Explain why it is not possible to use the method of repeated measures in all experiments.

10. Summarize the factors that researchers should consider before selecting an independent or correlated two-groups design to conduct their research. Discuss the advantages and disadvantages of these designs?

11. Define within-group and between-group variance. How are they related? Which of the two types of variance do researchers attempt to minimize? Discuss the factors that influence each of these variance sources?

12. Describe the general formula for statistical tests. What information is placed in the numerator and the denominator? Do researchers seek smaller or larger statistical test values? Explain your answer.

13. Discuss why researchers must use a larger critical t-value for a correlated-groups design than for an independent-groups design.

14. Define ex post facto research and summarize the benefits and limitations of this research method.

Answers for the multiple choice, matching, and true/false items

Multiple Choice		Matching		True and False	
1.	E	1.	B	1.	T
2.	A	2.	G	2.	T
3.	D	3.	C	3.	F
4.	C	4.	M	4.	F
5.	A	5.	H	5.	T
6.	B	6.	F	6.	T
7.	D	7.	J	7.	F
8.	A	8.	K	8.	F
9.	C	9.	D	9.	T
10.	B	10.	I	10.	F
11.	D	11.	E	11.	F
12.	D	12.	L	12.	T
13.	C	13.	A	13.	F
14.	A				
15.	C				
16.	B				
17.	C				
18.	E				
19.	A				
20.	B				
21.	A				
22.	E				
23.	D				
24.	B				
25.	C				
26.	D				
27.	A				
28.	B				

Chapter 11

Inferential Tests of Significance I: *t*-tests

Learning Objectives

By the end of this chapter you should be able to:

1. Summarize the rationale for using inferential statistics to make research decisions.

2. Describe the relation between experimental designs and statistical analysis.

3. List and explain the steps involved with determining if the results of an inferential statistical test is significant.

4. Compare and contrast independent and correlated participant samples.

5. Recognize that there are two different types of *t*-tests for two-group designs.

6. Calculate, interpret, and communicate the results of a *t*-test for independent samples.

7. Calculate, interpret, and communicate the results of a *t*-test for correlated samples.

8. Summarize the relation between *t*-tests and directional and non-directional hypotheses.

By the end of this chapter you should learn when and how to use the following formulas:

I. Formula and steps for calculating an independent samples *t*-test.

$$t = \frac{M_1 - M_2}{SE_{diff}}$$

(1)

Where:

M_1 = the mean of participant group 1
M_2 = the mean of participant group 2
SE_{diff} = the standard error of the difference between the means

1. The first step for conducting an independent samples t-test involves <u>calculating the means</u> for participant group 1(M_1) and participant group 2 (M_2).

2. Next, <u>calculate the variability index</u> or in this case the standard error of the difference between means (SE_{diff}). If your participant groups are equal (i.e., $N_1 = N_2$) then you would use formula 2 to calculate the SE_{diff}.

$$SE_{diff} = \sqrt{\left(\frac{\Sigma X_1^2 - \frac{(\Sigma X_1)^2}{N_1} + \Sigma X_2^2 - \frac{(\Sigma X_2)^2}{N_2}}{N_1(N_2 - 1)} \right)}$$

(2)

Where:

N_1	=	the number of participants in group 1
N_2	=	the number of participants in group 2
ΣX_1^2	=	the sum of the squared scores for group 1
ΣX_2^2	=	the sum of the squared scores for group 2
$(\Sigma X_1)^2$	=	the sum of the scores for group 1 squared
$(\Sigma X_2)^2$	=	the sum of the scores for group 2 squared

• If your participant <u>groups are unequal</u> ($N_1 \neq N_2$), then you would use formula 3 to calculate the SE_{diff}.

$$SE_{diff} = \sqrt{\left(\frac{\Sigma X_1^2 - \frac{(\Sigma X_1)^2}{N_1} + \Sigma X_2^2 - \frac{(\Sigma X_2)^2}{N_2}}{N_1 + N_2 - 2} \right) \left(\frac{1}{N_1} + \frac{1}{N_2} \right)}$$

(3)

3. Now that you have the value for the SE$_{\text{diff}}$, you can <u>calculate the t-statistic</u>. This requires subtracting the mean for participant group 1 (M_1) from the mean of participant group 2 (M_2) and then dividing this value by your appropriate SE$_{\text{diff}}$ (i.e., equal or unequal participant groups).

$$t = \frac{M_1 - M_2}{SE_{diff}}$$

4. Next, determine the <u>degrees of freedom</u> for the analysis. This can be accomplished by using formula 4.

$$df = (N_{group_1} - 1) + (N_{group_2} - 1) \qquad (4)$$

5. Lastly, <u>calculate the effect size</u> (Cohen's d) with formulas 5 and 6.

$$d = \frac{M_1 - M_2}{s} \qquad (5)$$

Where:

$$s = SE_{diff} \left(\sqrt{N_1} \right) \qquad (6)$$

II. <u>Formula and steps for calculating a correlated samples t-test</u>

$$t = \frac{\bar{X} - \bar{Y}}{s_{\bar{D}}} = \frac{\bar{X} - \bar{Y}}{\frac{s_D}{\sqrt{N}}} \qquad (7)$$

Where:

$\bar{X}$	=	the mean of participant group 1
$\bar{Y}$	=	the mean of participant group 2
$s_{\bar{D}}$	=	the standard error of the deviation scores
s_D	=	the standard error
N	=	the number of participants

1. The first step for conducting a correlated samples *t*-test involves <u>calculating the means</u> for group 1 ($\bar{X}$) and group 2($\bar{Y}$).

2. Next, calculate the <u>standard error of the deviation scores</u> ($s_{\bar{D}}$). However, before you can do this you need to calculate the standard error (s_D) given that the equation for the standard error of the deviation scores requires a standard error value. The standard error can be calculated using formula 8.

$$S_D = \sqrt{\frac{\Sigma D^2 - \frac{(\Sigma D)^2}{N}}{N-1}}$$

(8)

Where:

ΣD	=	the sum of the differences between each pair of scores
ΣD^2	=	the sum of the squared differences for each pair of scores
D	=	the difference between group 1 ($\bar{X}$) and group 2 ($\bar{Y}$)
N	=	the number of pairs of scores

- Once the <u>standard error</u> (s_D) is calculated, you can then calculate the <u>standard error of the deviation scores</u> ($s_{\bar{D}}$) using formula 9.

$$s_{\bar{D}} = \frac{s_D}{\sqrt{N}}$$

(9)

Where:

s_D	=	the standard error
N	=	the number of pairs of scores

3. Next, <u>calculate the *t*-test for correlated samples</u> using formula 7.

$$t = \frac{\bar{X} - \bar{Y}}{s_{\bar{D}}} = \frac{\bar{X} - \bar{Y}}{\frac{s_D}{\sqrt{N}}}$$

4. Then determine the <u>degrees of freedom</u> for the analysis using formula 10.

$$df = N_{number\ of\ pairs\ of\ participants} - 1$$

(10)

5. Lastly, calculate the <u>effect size</u> (Cohen's *d*) using formulas 11 and 12.

$$d = \frac{\overline{X}_1 - \overline{Y}_2}{s} \tag{11}$$

Where:

$$s = s_{\overline{D}} \left(\sqrt{N_1} \right) \tag{12}$$

Expanded Outline

I.   Inferential Statistics
        The Relation Between Experimental Design and Statistics
        The Logic of Significance Testing
            What is Significant?
                Null Hypothesis

            Samples and Populations
                Random Selection
                Random Assignment
        Analyzing Two-Group Designs
            t-test
            Interval-level Data
            Ratio-level Data
            Independent Groups

II.   The t-test for Independent Samples

        Calculating the t-value by Hand
            Standard Error (SE)
            Standard Deviation (SD)
            Degrees of Freedom (df)
        Statistical Issues
        Computer Analysis of a *t*-test for Independent Samples

            Descriptive Statistics
            Homogeneity of Variance
            Heterogeneity of Variance
            Robust

Translating Statistics into Words

APA Style

Practice Exam

Multiple Choice

Identify the letter of the choice that best completes the statement or answers the question.

_____ 1. Inferential statistics with small values occur _____ by chance; whereas inferential statistics with large values occur _____ by chance.

    a. never, always   c. frequently, rarely

    b. always, never   d. rarely, frequently

_____ 2. Which statistical test should be used to analyze the results of a study with one IV that has two levels or treatment conditions?

    a. z-test   c. t-test

    b. correlation   d. F-test

_____ 3. A t-test for independent samples is an appropriate statistical test for

    a. randomly assigned participants.   c. matched pairs.

    b. natural pairs.   d. repeated measures experiments.

_____ 4. The t-test is an inferential statistical test used to evaluate the difference between the means of _____ groups.

    a. 2   c. 4

    b. 3   d. 5

_____ 5. The following formula is used to calculate a(n)

$$t = \frac{\bar{X} - \bar{Y}}{s_{\bar{D}}} = \frac{\bar{X} - \bar{Y}}{\dfrac{s_D}{\sqrt{N}}}$$

    a. correlated samples t-test.   c. standard error of the difference.

    b. independent samples t-test.   d. ANOVA.

_____ 6. Which measure of variability or variability index is needed to calculate a t-test for independent samples?

    a. variance   c. SE_{DIFF}

    b. SEM   d. range

_____ 7. When using t-tests for two independent groups, the degrees of freedom are equal to _____ ; whereas when using t-tests for two correlated groups, the degrees of freedom are equal to _____ .

a. $N - 1, N - 2$
b. $N - 2, N - 1$
c. $N + 1, N + 2$
d. $N + 2, N + 1$

_____ 8. Compared to the t-test for correlated groups, the t-test for independent groups has _____ degrees of freedom.

a. more
b. less
c. equal
d. not enough information

_____ 9. Which of the following participant factors should researchers consider when calculating the standard error of the difference (SE_{DIFF})?

a. ethnicity
b. gender
c. region of the United States
d. size

_____ 10. A(n) _____ statistic is used to test the assumption of homogeneity of variance.

a. Cohen's d
b. p-value
c. F_{max}
d. SE_{DIFF}

_____ 11. When the variability of the scores of two groups is <u>not</u> similar, researchers refer to this as

a. homogeneity of variance.
b. heterogeneity of variance.
c. robustness.
d. both a and b
e. both b and c

_____ 12. Professor Parker has tested the assumption of homogeneity of variance for his two groups. What should he conclude given that the result of his F_{max} test was:

$F_{max} = 1.729, p = 0.26$

a. the group variances are different
b. heterogeneity of variance
c. homogeneity of variance
d. the results are inconclusive

_____ 13. A(n) _____ statistical test that can tolerate violations of its assumptions (i.e., homogeneity of variances) and still provide accurate results is called

a. robust
b. reliable
c. descriptive
d. unbiased

_____ 14. You are provided with the following results of a t-test for independent groups. Use this information to answer questions 14-17.

$$F_{max} \text{ test:} \quad F = 1.634 \quad p = 0.222$$

$$t(24) = 3.37, p = .049$$

$$\text{Cohen's } d = 0.85$$

What can you conclude from the result of the F_{max} test?

a. the variances of the groups were not equal   c. the F_{max} test was significant
b. the variances of the groups were equal   d. the results were inconclusive

_____ 15. How many degrees of freedom were used for this analysis?

a. 3.37   c. 24
b. 1.64   d. .049

_____ 16. The results of the t-test analysis were

a. significant at the .001 level.   c. significant at the .05 level.
b. significant at the .01 level.   d. not significant.

_____ 17. The effect size for this analysis was

a. small.   c. large.
b. moderate.   d. inconclusive.

_____ 18. When communicating the statistical results of a study researchers use the

a. APA Publication Manual.   c. APA Research Ethics Manual.
b. Psychological Thesaurus.   d. Diagnostic Statistics Manual.

_____ 19. The measure of variability used for calculating a correlated samples t-test is the

a. median.   c. standard error of the difference.
b. variance.   d. standard error of the deviation scores.

_____ 20. Researchers seek high correlation coefficients when comparing the relations between

a. independent participant samples.   d. both a and b
b. matched participant pairs.   e. both b and c
c. the same participant used for repeated measures.

_____ 21. The one-tailed *t*-test evaluates the probability of _____ outcome(s); whereas the two-tailed *t*-test evaluates the probability of _____ outcome(s).

     a. multiple, one   c. both possible, one
     b. one, more than two   d. one, both possible

Matching

a. $N - 1$

b. *df*

c. $t(27) = 2.61, p = .04$

d. Cohen's *d*

e. standard error of the deviation scores

f. independent groups

g. F_{max}

h. standard error of difference between means

i. $t(13) = 2.37, p = .06$

j. a robust test

k. $N - 2$

l. correlated groups

_____ 1. a significant *t*-test finding using an alpha level of .05

_____ 2. a measure of variability used for calculating a correlated samples *t*-test

_____ 3. the symbol for degrees of freedom

_____ 4. a statistic used to test the assumption of homogeneity of variance

_____ 5. a non-significant *t*-test finding using an alpha level of .05

_____ 6. a measure of variability used for calculating an independent samples *t*-test

_____ 7. randomly assigned participant groups

_____ 8. used to calculate the degrees of freedom for two independent groups

_____ 9. a measure of effect size

_____ 10. can tolerate violations of its assumptions and still provide accurate answers

_____ 11. matched participant groups

_____ 12. used to calculate the degrees of freedom for two correlated groups

True/False

Indicate whether the sentence or statement is true or false.

_____ 1. Researchers should determine their experimental design after collecting data.

_____ 2. Decisions associated with using a certain *t*-test are based on methods used to assign participants to groups.

_____ 3. The sign (positive or negative) of a *t*-test is inconsequential when using a two-tailed *t*-test.

_____ 4. A *t*-test is not robust with regard to the assumption of homogeneity.

_____ 5. The formula for calculating a correlated samples *t*-test is equivalent to the formula used for calculating an independent samples *t*-test.

_____ 6. Researchers want the relation between matched pairs to be highly correlated to increase statistical control.

_____ 7. When using a one-tailed *t*-test the sign (positive or negative) of the resulting *t*-value is not critical.

_____ 8. Researchers typically set the alpha level at .01 because it places the probability of Type I and Type II errors at acceptable levels.

_____ 9. Compared to the *t*-test for independent samples, the *t*-test for correlated groups has fewer degrees of freedom.

_____ 10. A *t*-test is a descriptive statistical test used to evaluate the difference between two means.

_____ 11. Homogeneity of variance is a term for communicating that the variances of the participant groups are not equal.

Short Answer

1.	Summarize the logic involved with using inferential statistics to make decisions about the results of a study.

2.	Describe the differences between the formulas for calculating the standard error of the difference between means (i.e., SE_{diff}) for an independent samples t-test with equal (i.e., $N_1 = N_2$) and unequal ($N_1 \neq N_2$) participant groups.

3.	Explain why researchers must use the standard error of the difference (SE_{diff}) instead of the standard error of the mean (SEM) when calculating an independent samples t-test.

4.	In the computer analysis of t-tests for independent samples section of the book the authors used the saying, "garbage in, garbage out." Explain the significance of this statement and provide one method for avoiding this problem.

5.	Define the terms homogeneity and heterogeneity of variance. Discuss how these terms are related to a t-test. In what situations are t-tests robust?

6.	Summarize the rationale behind researchers seeking strong correlation coefficients for matched participant samples instead of independent participant samples.

7.	Describe how directional and non-directional hypotheses are related to t-tests.

8.	Explain why the natural pairs research method can be viewed as a mixture of matched pairs and repeated measures.

9.	Discuss why it is important for researchers to pay attention to the sign (i.e., positive or negative) of an obtained t-value for a one-tailed t-test (directional hypothesis), but not for a two-tailed t-test (non-directional).

10.	Identify and describe a procedure (as discussed in chapter 11) for selecting a one-tailed or directional t-test. Explain the rationale for using this procedure.

SPSS Computer Practice Problem 11.1 Using SPSS 11.0 for Windows

Basically when conducting a statistical analysis using SPSS for Windows you will name your variables, enter your data, and finally analyze your data by selecting options from a toolbar. In the following section a step by step procedure of an example problem will be provided to guide you in calculating a *t*-test for independent samples. By following the systematic instructions provided and by referring to the screen figures (SF) when prompted, you should be able to independently conduct a *t*-test for independent samples.

Data

The fictional data for this problem is based on a sports psychologist who was interested in studying the effect that fatigue has on the accuracy of dart players. He randomly assigned 14 players from a dart league to either an experimental ($n = 7$) or a control group ($n = 7$). The accuracy task involved the players throwing darts at the bull's-eye on a dart board. Each player threw 15 darts and each bull's-eye was awarded one point. The control group was immediately tested. The experimental group was tested after being deprived of sleep for 24 hours. The sports psychologist wanted to determine if the accuracy scores between the two groups would differ. The raw data for this analysis is provided in Table 1.

Table 1

Participant Groups	
Control (group 2) (non-sleep deprived)	**Experimental (group 1)** (sleep deprived)
9	2
12	8
8	3
7	9
12	4
8	2
7	7

While working on this example problem consider the following questions:

1. How will you express the null and alternative or experimental hypotheses?

2. Is this a directional or non-directional hypothesis?

3. What type of *t*-test analysis should be used?

4. Will the results be significant at the pre-set .05 alpha level?

5. Will the variances for the two groups be homogenous (i.e., equal) or heterogeneous (i.e., unequal)? How will you reach this conclusion?

6. How will you calculate the degrees of freedom for this analysis?

7. How will you write the statistical results for this experiment?

8. How will you interpret the results in non-statistical language?

Procedure for Completing Example Problem 11.1

1. **Data Entry.** This example will have data for a control group (non-sleep deprived) and an experimental group (sleep deprived). For the purpose of this example the variable names will be "group" and "score." The variable named "group" will include both participant groups, however each group will be coded differently so that SPSS recognizes they are separate groups. For example, the participants in the experimental group will be coded with a "1" and the participants in the control group with a "2" (see SF-11.1). The variable named "score" will include corresponding information about the accuracy of each participant or how many bull's-eyes they hit. For example, the first participant in screen figure 11.1, row one, column one, will be from the experimental group ("1") and their corresponding accuracy score will be a "2." The data for this analysis should resemble screen figure 11.1.

SF-11.1. Entering data to conduct a *t*-test for independent samples.

Before conducting the *t*-test for independent samples analysis it may be helpful to label the values you have used in your "group" variable. This will make your output much easier to interpret. Labeling these values can be accomplished by selecting the "Variable View" button on the bottom left corner of the data entry screen (see SF-11.1). Once selected your screen should resemble screen figure 11.2. Next click on the box below the "Values" column for row one. This box should have the word "None" in it before you select it. Once selected a "Value Labels" command box will appear (see SF-11.2). At this point enter a "1" in the "Value" cell and enter the name ("experimental") of this value in the "Value Label" cell. Then select the "Add" button in the middle left hand corner of the "Value Labels" command box to move this information into the larger white box. If you make a mistake you can use the "Remove" command and start over. Repeat this process for the control group except enter a "2" and name this value "control." Once you have finished select the "OK" button located in the upper right hand corner of the "Value Labels" command box and return to the data entry screen by selecting "Data View" at the bottom left hand corner of the screen (see SF-11.2).

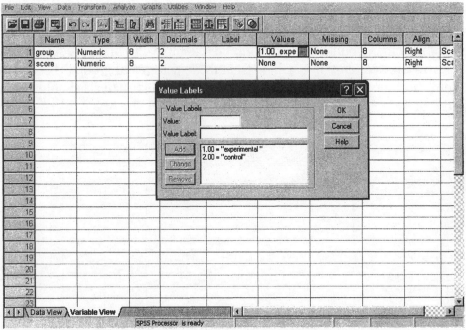

SF-11.2. Entering data: Labeling the coded values.

2. **Conducting the *t*-test for independent samples analysis.** In order to conduct a statistical analysis using SPSS you will need to select the "Analyze" toolbar option at the top of the screen. A dropdown box will appear providing further data analysis options. Select the "Compare Means" option. After selecting this option SPSS will provide you with further analysis alternatives to the right of the initial dropdown box (see SF-11.3). At this point select the "Independent-Samples T Test…" option.

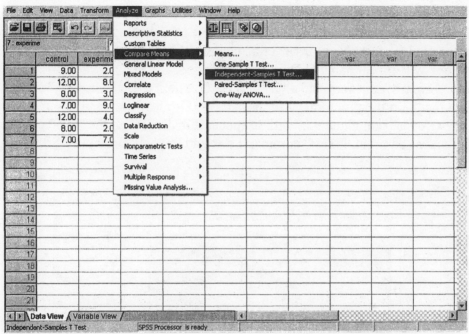

SF-11.3. Selecting the *t*-test for independent samples analysis.

Once you have selected the "Independent-Samples T Test…" option you should be able to view a command box named "Independent-Samples T Test" (see SF-11.4). At this point the variables "score" and "group" will appear in the large white box on the left hand side of the "Independent-Samples T Test" command box. Move the "score" variable into the "Test Variable(s)" box by first highlighting (it should appear blue) this variable and then by selecting the arrow in the middle of the screen (see SF-11.4). Next, move the "group" variable into the "Grouping Variable" box by selecting the arrow button closest to the "Grouping Variable" box. At this point you will need to define your groups or categories for the "group" variable. This can be done by selecting the "Define Groups" button in the "Independent-Samples T Test" command box. Once selected another command box will appear named "Define Groups" (see SF-11.4). Type a "1" into the box next to the "Group 1" category and a "2" into the box next to the "Group 2" category. Now select the "Continue" command in the "Define Groups" command box. At this point the "Define Groups" command box will disappear. Finally, select the "OK" command in the "Independent-Samples T Test" command box and this will prompt SPSS to run your analysis.

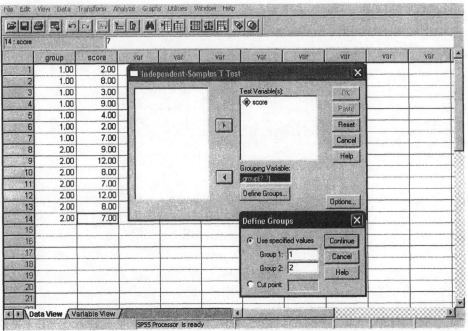

SF-11.4. Conducting the *t*-test for independent samples analysis: Defining groups.

152

3. **Reviewing the Output for the Independent-Samples *t*-test Analysis.** Your output should resemble the output in SF-11.5. If it does not you may want to check your raw data for data entry errors. The output in Screen Figure 11.5 provides descriptive and inferential information. The first output labeled "Group Statistics," provides the levels of the grouping variables (i.e., experimental or control), the number of cases, the means and standard deviations, and the standard error of the mean or the standard deviation of the distribution of means for each grouping. The second output box provides information about Levene's Test for Equality of Variances, which is a significance test of the null hypothesis that the variances of the two groups are equal or homogeneous. This is an important assumption of the *t*-test for independent means. If this test is significant at the .05 alpha level researchers do not assume that the variances of the two populations are equal (i.e., they are heterogeneous). Conversely, if this test is not significant, researchers can assume that the variances of the two populations are equal (i.e., they are homogeneous). You should recognize that SPSS provides two values for the obtained *t*-test statistic, one for equal variances assumed and one for equal variances not assumed. The second output box also provides the degrees of freedom and the probability, or exact chance, of obtaining this *t*-score on this particular distribution. Lastly, the second output box also provides the standard error of the differences and a 95 percent confidence interval.

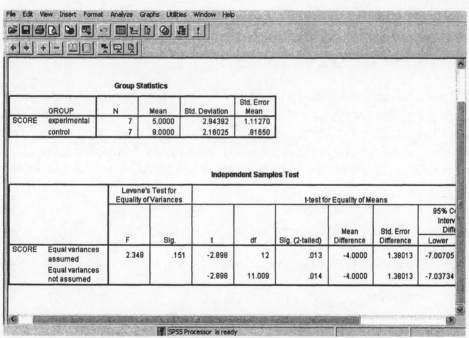

SF-11.5. Output for the independent-samples t-test analysis.

4. **Interpreting the Results**.

<u>How would you express the null and alternative or experimental hypotheses?</u>

The null hypothesis can be expressed as (H_o: $M_1 = M_2$)
The alternative hypothesis can be expressed as (H_a: $M_1 \neq M_2$)

<u>Was this a directional or non-directional hypothesis?</u>

This was a non-directional hypothesis because the researcher was only interested in learning if the means between the control and experimental groups would differ.

<u>What type of *t*-test analysis was used?</u>

This was a *t*-test for independent samples because the participant groups were formed by random assignment. Also the control group had no relation to, or effect on, the experimental group.

<u>Was the result significant at the .05 alpha level?</u>

Yes the result was significant given that the *p*-value (i.e., .013) was less than the pre-set significance level of .05.

<u>Were the variances of the two groups homogenous (i.e., equal) or heterogeneous (i.e., unequal)? How did you reach this conclusion?</u>

In this example the results of the Levene's test for equality of variances was <u>not</u> significant given that the *p*-value of .151 was greater than .05. Consequently, the researcher has no reason to doubt the assumption of equal population variances and can assume the group variances are homogeneous. As a result the researcher would utilize the *t*-test statistic that pertains to the "Equal variances assumed" row in the SPSS output (see SF-11.5).

<u>What were the degrees of freedom for this analysis? How did you reach this conclusion?</u>

The degrees of freedom can be calculated using the formula for independent samples:

$$df = (N_{experimental} - 1) + (N_{control} - 1)$$

$$df = (7 - 1) + (7 - 1)$$

$$df = 12$$

<u>How could you write the statistical results for this experiment?</u>

The dart players in the control group or non-sleep deprived condition ($M = 9.0$, $SD = 2.16$) hit significantly more bull's-eyes [$t(12) = -2.89$, $p = .013$] than did the players in the experimental or sleep deprived group ($M = 5.0$, $SD = 2.94$).

How could you interpret the results in non-statistical language?

Fatigue significantly decreased the performance of the dart players who were sleep deprived when compared with the dart players who were not sleep deprived.

Answers for the multiple choice, matching, and true/false items

Multiple Choice		Matching		True and False	
1.	C	1.	C	1.	F
2.	C	2.	E	2.	T
3.	A	3.	B	3.	T
4.	A	4.	G	4.	F
5.	A	5.	I	5.	F
6.	C	6.	H	6.	T
7.	B	7.	F	7.	F
8.	A	8.	K	8.	F
9.	D	9.	D	9.	T
10.	C	10.	J	10.	F
11.	B	11.	L	11.	F
12.	C	12.	A		
13.	A				
14.	B				
15.	C				
16.	C				
17.	C				
18.	A				
19.	D				
20.	E				
21.	D				

Chapter 12

Designing and Conducting Experiments with More than Two Groups

Learning Objectives

By the end of this chapter you should be able to:

1.   Summarize the factors for designing and conducting experiments with more than two groups.

2.   Explain why one-IV experiments are not inferior to those that use two or more IVs.

3.   Discuss why a research question should dictate the type of experimental design used to conduct a study.

4.   List and summarize the relevant questions that researchers should consider when conducting experiments using a multiple-groups designs.

5.   Describe the distinguishing characteristics between the two-groups and multiple-groups experimental designs.

6.   Discuss the relevant questions that researchers should consider before determining whether to use a two-groups design or a multiple-groups design.

7.   Explain several factors that researchers should consider before deciding on whether to use a multiple-groups design with independent or correlated samples.

8.   Describe the participant group limitations associated with using a multiple-groups design.

9.   Recreate the general formula for a statistical test.

10.   Summarize why practical considerations in dealing with research participants makes the multiple-correlated groups design considerably more complicated than the multiple-independent groups design.

Expanded Outline

I. Experimental Design: Adding to the Basic Building Block
    Experimental Design
    Multiple-Group Design
    How Many IVs?
    Principle of Parsimony
    How Many Groups?
    Levels

    Treatment Groups

    Assigning Participants to Groups
    Independent Samples
    Random Assignment
    Control Procedure
    Confounded Experiment
    Correlated Samples
    Matched Sets
    Matching Variable
    Repeated Measures

    Natural Sets

II. Review Summary

III. Check Your Progress

IV. Deciding on an Experimental Design
    Comparing the Multiple-Group and Two-Group Designs

    Comparing Multiple-Group Designs
    Choosing a Multiple-Group Design
    Control Issues
    Practical Considerations
    Variations on the Multiple-Group Design
    Comparing Different Amounts of an IV
    Placebo Effect
    Dealing with Measured IVs
    Ex Post Facto Research

V. Review Summary

157

Practice Exam

Multiple Choice

Identify the letter of the choice that best completes the statement or answers the question.

_____ 1.   The most common type of two-group design uses _____ and _____ groups.

a. experimental, natural   c. control, experimental
b. experimental, correlated   d. control, confounder

_____ 2.   An experiment with one IV and three levels requires a

a. single-group design.   c. multiple-groups design.
b. two-groups design.   d. control-group design.

_____ 3.   When using experimental designs that are composed of two or more participant groups, the IV levels refer to the

a. number of experimental groups.   c. number of IVs.
b. different amounts or types of the IV.   d. number of DVs.

_____ 4.   **Use the following information to answer questions 4-8.** Annie is studying a new nutritional supplement for puppies. She wants to determine whether this new supplement will affect their performance on physical and cognitive tasks. She decides to use the following three dietary treatment conditions:

Group 1: puppies only receive the nutritional supplement
Group 2: puppies receive a 50 percent mixture of the nutritional supplement and the common dog food.
Group 3: puppies only receive the common dog food.

What type of experimental design will she likely use?

a. single-group   c. multiple-groups
b. two-groups   d. control-group

_____ 5.   How many IVs will she use in her study?

   a. 0   c. 2
   b. 1   d. 3

_____ 6.   How many levels will she use in her study?

   a. 1   c. 3
   b. 2   d. 4

_____ 7.   What is/are the experimental condition(s) that she will be using in her study?

   a. receiving only the nutritional supplement   d. both a and b
   b. receiving the 50 percent mixture   e. both b and c
   c. receiving only the common dog food

_____ 8.   What is/are the control condition(s) that she will be using in her study?

   a. receiving only the nutritional supplement   d. both a and b
   b. receiving the 50 percent mixture   e. both b and c
   c. receiving only the common dog food

_____ 9.   In correlated-group designs researchers compare differences _____ groups; whereas in independent-group designs researchers compare differences _____ differing groups of participants.

   a. within, between   b. between, within

_____ 10. Sadie has conducted a thorough review of the literature pertaining to her study to determine if her IV has had an affect in other studies. She is not able to find any evidence that would support that her IV has had an affect in other studies. What type of experimental design should she choose given these circumstances?

   a. control-group   c. two-groups (presence-absence)
   b. single-group   d. multiple-groups

_____ 11. The limit for the number of groups used in a multiple-groups experiment is

   a. 3.   c. 5.
   b. 4.   d. based only on practical considerations.

____ 12. A researcher's decision to use the multiple-independent-groups design versus the multiple-correlated-groups design revolves around

    a. control issues.   c. bias issues.
    b. reliability issues.   d. the Principle of Parsimony.

____ 13. The multiple-independent-groups design uses the control technique of

    a. matching.   d. natural pairs.
    b. random assignment of participants.   e. all of the above
    c. repeated measures.

____ 14. Multiple-correlated-group designs use the control technique of

    a. matching.   c. natural pairs.
    b. repeated measures.   d. all of the above

____ 15. The multiple-correlated group designs use control techniques to

    a. assure equal participant groups.   d. both a and b
    b. reduce error variability.   e. both b and c
    c. assure random participant assignment.

____ 16. Which of the following represents the general formula for a statistics test?

    a. within-group variability/   c. error variability/
       error variability   within-group variability
    b. between-groups variability/   d. error variability/
       error variability   between-groups variability

____ 17. An experimental effect that is due to expectation or suggestion rather than the IV is referred to as a(n)

    a. valid.   c. placebo effect.
    b. ex post facto study.   d. a nuisance variable.

____ 18. Using a measured IV instead of a manipulated IV is associated with a(n)

    a. ex post facto study.   c. placebo factor.
    b. confounder.   d. nuisance variable.

_____ 19. Justine is conducting research with chimpanzees at the zoo. She measures their dietary preferences on three separate occasions. This is an example of a(n) _____ experiment.

a. repeated measures
b. matched pairs

c. natural sets
d. correlation

_____ 20. _____ group designs provide extra advantages for experimental control relative to _____ group designs.

a. Multiple-independent, multiple-correlated
b. Multiple-independent, ex post facto

c. Multiple-correlated, ex post facto
d. Multiple-correlated, multiple-independent

Matching

a. placebo effect
b. treatment group
c. error variability
d. control group
e. independent group
f. ex post facto research

g. random assignment
h. experimental design
i. between-groups variability
j. correlated group
k. random assignment
l. treatment conditions

_____ 1. a value placed in the numerator of the general formula for a statistical test

_____ 2. participant group formed by random assignment

_____ 3. plan for selecting and assigning participants to experimental conditions, controlling extraneous variables, and gathering data.

_____ 4. related participants

_____ 5. a method where the experimenter can only classify, categorize, or measure, the IV

_____ 6. an experimenter effect due to expectation or suggestion rather than the IV

_____ 7. procedure in which participants have an equal chance of placement in any group

_____ 8. a procedure used to control for potential extraneous variables

_____ 9. participants who do not receive the IV

_____ 10. a value placed in the denominator of the general formula for a statistical test

_____ 11. participants who receive the IV

_____ 12. differing amounts or types of an IV

True/False

Indicate whether the sentence or statement is true or false.

_____ 1. The two-groups design is the most basic experimental design.

_____ 2. Research using one IV is typically inferior to research using two or more IVs.

_____ 3. A multiple-groups design cannot have a control group.

_____ 4. Changing a two-groups design into a multiple-groups design only requires adding another level to your IV.

_____ 5. Your research question should dictate which experimental design you choose for your study.

_____ 6. Increasing the error variability in the denominator of the general formula for a statistical test will result in a larger computed statistical value.

_____ 7. Using a correlated-groups design reduces your degrees of freedom.

_____ 8. Correlated designs often produce weaker tests for yielding statistical significance.

_____ 9. It is not possible to use measured IVs in a multiple-groups design.

_____ 10. When using an ex post facto research design researchers manipulate the IV.

Short Answer

1. Summarize why one-IV studies are not inferior to studies using two or more IVs.

2. Discuss why your research question should dictate the type of experimental design you use in your study.

3. Summarize the questions that researchers should consider when conducting an experiment using a multiple-group design (Hint: refer to the multiple-group design section of this chapter).

4. Describe the distinguishing characteristics between the two-groups and multiple-groups experimental designs. Explain how these differences are related to a blueprint for constructing a building.

5. Explain several factors that you should consider before deciding on whether to use a multiple-groups design with independent or correlated samples. What are the advantages and disadvantages of each design?

6. Provide an example of a research question that could be answered using a multiple-groups design. Provide an example of a research question that could be answered using a two-groups design. Are there instances where an experimental question could be answered with a multiple-groups design and not a two-groups design? Why or why not?

7. List several questions you should consider when faced with a situation in which either a two-groups design or a multiple-groups design is appropriate.

8. Describe the participant group limitations associated with using a multiple-groups design.

9. Recreate the general formula for a statistical test. Discuss what will happen if you reduce the error variability in the denominator of this equation. How will this affect the null hypothesis?

10. Explain why practical considerations in dealing with research participants makes the multiple-correlated groups design considerably more complicated than the multiple-independent groups design.

Answers for the multiple choice, matching, and true/false items

Multiple Choice		Matching		True and False	
1.	C	1.	I	1.	T
2.	C	2.	E	2.	F
3.	B	3.	H	3.	F
4.	C	4.	J	4.	T
5.	B	5.	F	5.	T
6.	C	6.	A	6.	F
7.	D	7.	K	7.	T
8.	C	8.	G	8.	F
9.	A	9.	D	9.	F
10.	C	10.	C	10.	F
11.	D	11.	B		
12.	A	12.	L		
13.	B				
14.	D				
15.	D				
16.	B				
17.	C				
18.	A				
19.	A				
20.	D				

Chapter 13

Analyzing and Interpreting Experiments with More Than Two Groups

Learning Objectives

By the end of this chapter you should be able to:

1. Analyze and interpret experiments with more than two groups.

2. Determine when it is appropriate to use a two-group or multiple-group research design.

3. Summarize how the two-group and multiple-group designs are similar and different.

4. Discuss the appropriate statistical analysis procedure for multiple-group designs.

5. List and describe the necessary steps for calculating and interpreting a one-way ANOVA for independent and correlated samples.

6. Interpret and communicate (using APA format) computer output for a one-way ANOVA for independent and correlated samples.

7. Recreate the general conceptual formula for the multiple-group design.

8. Describe and interpret the information found in a source table.

9. Define and explain what information is needed to calculate a mean square and an F-ratio.

10. Summarize why there is an additional step involved with calculating a one-way ANOVA for correlated samples.

11. Explain the purpose for using post hoc tests.

By the end of this chapter you should learn when and how to use the following formulas:

I.   <u>General Conceptual Formula for an ANOVA</u>

$$F = \frac{Between - Groups \quad Variability}{Within - Groups \quad Variability}$$

(1)

II.   Steps for Calculating a One-Way ANOVA for Independent Groups

1. Calculate the sum of scores for each group.
2. Sum the scores for all groups.
3. Square each individual score and sum those squares by group.
4. Sum the squared scores for all groups.
5. Calculate the total sum of squares using formula 2.

$$SS_{tot} = \Sigma X^2 - \frac{(\Sigma X)^2}{N}$$
(2)

6. Calculate the treatment sum of squares or between-group variability using formula 3.

$$SS_{between} = \Sigma \left[\frac{(\Sigma X^2)}{n} \right] - \left[\frac{(\Sigma X)^2}{N} \right]$$
(3)

Where:

n   =   number of scores in each column
N   =   total number of scores for all groups

7. Calculate the error sum of squares or the within-group variability using formula 4.

$$SS_{within} = \Sigma \left[\Sigma X^2 - \frac{(\Sigma X)^2}{N} \right]$$
(4)

8. Calculate the degrees of freedom (*df*) for each of the sums of squares using the following formulas:

df_{tot} = $N - 1$
$df_{between}$ = $K - 1$
df_{within} = $N - K$

Where:

N =   the total number of scores for all groups
K =   the number of IV levels or factors

9. Calculate the <u>mean squares</u> (MS) by dividing the sum of squares for our IV (between-groups variability) and error (within-groups variability) by their respective *df*. This can be accomplished by using formulas 5 and 6.

$$MS_{between} = \frac{SS_{between}}{df_{between}}$$

(5)

$$MS_{within} = \frac{SS_{within}}{df_{within}}$$

(6)

10. Calculate the *F*-obtained ratio by dividing the MS of the IV (MS between) by the MS for the error term (MS within). This can be accomplished by using formula 7.

$$F_{obt} = \frac{MS_{between}}{MS_{within}}$$

(7)

III.   <u>Steps for Calculating a One-Way ANOVA for Correlated Groups</u>

1. Calculate the sum of scores for each group.
2. Sum the scores for all groups.
3. Square each score and sum those squares by group.
4. Sum of the squared scores for all groups.
5. Calculate the <u>total sum of squares</u> using formula 1.

$$SS_{tot} = \Sigma X^2 - \frac{(\Sigma X)^2}{N}$$

(1)

6. Calculate the <u>treatment sum of squares</u> or between-groups variability by using formula 2.

$$SS_{between} = \Sigma \left[\frac{(\Sigma X^2)}{n} \right] - \left[\frac{(\Sigma X)^2}{N} \right]$$

(2)

Where:

n =   the number of scores in each column
N =   the total number of scores for all groups

7. Calculate the <u>participant sum of squares</u> using formula 3.

$$SS_{participants} = \Sigma \left[\frac{(\Sigma P^2)}{n} \right] - \left[\frac{(\Sigma X)^2}{N} \right]$$

(3)

Where:

n = the number of participant groups
ΣP = the sum of the participant scores for each row

8. Calculate the <u>error sum of squares</u> or the within-group variability by using formula 4.

$$SS_{within} = SS_{tot} - SS_{between} - SS_{participants}$$

(4)

9. Calculate the <u>df</u> for each of the SS by using the following formulas:

$$df_{tot} = N - 1$$

$$df_{participants} = n_{participants} - 1$$

$$df_{between} = n_{between} - 1$$

$$df_{within} = (n_{participants} - 1)(n_{between} - 1)$$

10. Calculate the <u>MS</u> by using the formulas 5 and 6.

$$MS_{between} = \frac{SS_{between}}{df_{between}}$$

(5)

$$MS_{within} = \frac{SS_{within}}{df_{within}}$$

(6)

11. Calculate the obtained <u>F-ratio</u> by using formula 7.

$$F_{obt} = \frac{MS_{between}}{MS_{within}}$$

(7)

Expanded Outline

I.   Before Your Statistical Analysis
          Analyzing Multiple-Group Designs
               One-way ANOVA
               One-way ANOVA for Independent Groups
               One-way ANOVA for Correlated Groups
          Planning Your Experiment

               Operational Definitions
          Rationale of ANOVA
               Between-Groups Variability
               Within-Groups Variability

II.   One-Way ANOVA for Independent Groups
          Calculating ANOVA by Hand

          Sum of Squares
          Total Sum of Squares
          Treatment Sum of Squares
          Error Sum of Squares
          Mean Square
               Degrees of Freedom (*df*)
               Variance
               F-ratio

               Post Hoc Comparisons
               Source Table

          Computer Analysis of ANOVA for Independent Groups
               Interpreting Computer Statistical Output
               Translating Statistics into Words

III.   Review Summary

IV.   Check Your Progress

V.   One-Way ANOVA for Correlated Groups

               Calculating ANOVA by Hand
                    Total Sum of Squares
                    Treatment Sum of Squares

Participant Sum of Squares
Error Sum of Squares
Mean Squares
F-ratio

Post Hoc Comparisons
Source Table

Computer Analysis of ANOVA for Correlated Groups
Interpreting Computer Statistical Output
Asymptotic
Translating Statistics into Words

VI.   The Continuing Research Problem

VII.   Review Summary

VIII.   Check Your Progress

IX.   Exercises

X.   Key Terms

XI.   Looking Ahead

Practice Exam

Multiple Choice

Identify the letter of the choice that best completes the statement or answers the question.

_____ 1.   The appropriate statistical analysis procedure for multiple-group designs is

    a. correlation.   c. *t*-test.
    b. *z*-test.   d. analysis of variance.

_____ 2.   Researchers who randomly assign their participants to comparison groups will analyze their data with a(n)

    a. *t*-test.   c. one-way ANOVA for independent groups.
    b. F_{max} test.   d. one-way ANOVA for correlated groups.

_____ 3. The effect of an IV is contained in

    a. within-group variability.   c. nuisance variables.
    b. between-group variability.   d. confounders.

_____ 4. When an IV has no effect or only a small effect, the _F_-obtained ratio will be approximately:

    a. -1.00   c. 1.00
    b. 0.00   d. 10.00

_____ 5. The _F_-obtained ratio is conceptually expressed as variability due to the _____ divided by variability due to the _____.

    a. IV, DV   c. IV plus error, error
    b. DV, IV   d. DV plus error, error

_____ 6. **Use the following information to answer questions 6-8.** Martha used 60 participants in her study. She randomly placed her participants into three equal groups with each participant group receiving a different level of the IV. How should she calculate the total degrees of freedom (_df_) for her study?

    a. (total # of scores) - 1   c. (total # of scores) - (# of IV levels)
    b. (# of IV levels) - 1   d. none of the above

_____ 7. How should Martha calculate the _df_ for her error term or within-group variability?

    a. (total # of scores) - 1   c. (total # of scores) - (# of IV levels)
    b. (# of IV levels) - 1   d. none of the above

_____ 8. How should Martha calculate the _df_ for her treatment effect or between-group variability?

    a. (total # of scores) - 1   c. (total # of scores) - (# of IV levels)
    b. (# of IV levels) - 1   d. none of the above

_____ 9. The statistical output for an ANOVA is typically presented in the form of a(n) _____ table.

    a. error   c. _t_
    b. _df_   d. source

____ 10. When conducting an ANOVA a(n) _____ is analogous to an estimate of the variance.

    a. mean square   c. F_{max}
    b. df d. t-test

____ 11. Researchers need _____ separate degrees of freedom values to obtain an F-critical value from the F-table.

    a. 2   c. 4
    b. 3   d. 5

____ 12. To obtain a mean square (MS), the sum of squares (SS) is divided by its

    a. df.   c. F-ratio.
    b. N.   d. p-value.

____ 13. Which of the following is used to calculate the within-groups df when conducting a one-way ANOVA for independent groups?

    a. N - 1   c. N - K
    b. K - 1   d. K - N

____ 14. Which of the following is used to calculate the between-groups df when conducting a one-way ANOVA for independent groups?

    a. N - 1   c. N - K
    b. K - 1   d. K - N

____ 15. Post hoc comparisons are used to determine significant differences between mean scores in

    a. single-group designs.   c. multiple-group designs.
    b. two-group designs.   d. none of the above.

____ 16. Which of the following is a post hoc test?

    a. Cohen's d c. Levene's Test
    b. F_{max} d. Tukey HSD

____ 17. Professor Smith is conducting a post hoc comparison. At this point he can assume that he has obtained a

a. non-significant result in a two-group experiment.
b. non-significant result in a multiple-group experiment.
c. significant result in a two-group experiment.
d. significant result in a multiple-group experiment.

____ 18. The _____ sum of squares (SS) are used to represent the total variability of the DV in the experiment.

a. total
b. within-group
c. between-group
d. mean

____ 19. To obtain the F-ratio researchers divide the _____ by the _____.

a. between-groups SS, within-groups SS
b. between-groups df, within-groups df
c. between-groups MS, within-groups MS
d. between-groups p-value, within-groups p-value

____ 20. You are reading about the results of a one-way ANOVA for independent groups with three treatment conditions in the *Psi-Chi Journal of Undergraduate Research*. The result is written as:

$F(2, 21) = 4.71, p = .02$.

How many total participants were used in this study?

a. 2
b. 3
c. 21
d. 24

____ 21. How many IV levels were used in question 20?

a. 2
b. 3
c. 4
d. 5

____ 22. In a one-way ANOVA for independent groups researchers calculate the SS for the

a. total source.
b. between-groups source.
c. the within-groups source.
d. all of the above

_____ 23. In a one-way ANOVA for correlated groups researchers calculate the SS for the

    a. total source.   d. participant source.
    b. between-groups source.   e. all of the above
    c. within-groups source.

_____ 24. When comparing the results of a one-way ANOVA for independent groups with the results of a one-way ANOVA for correlated groups, researchers typically find that

    a. there are fewer _df_ for the error   d. all of the above
        term in the correlated groups.
    b. the _F_-value for the correlated   e. none of the above
        groups test is larger.
    c. the proportion of variance accounted
        for is often larger with correlated groups.

_____ 25. Samuel wants to compare the results of three different personality tests and he obtains a random sample of participants with each group taking a different test. What type of analysis should he use?

    a. a one-way ANOVA for   c. a _t_-test for independent groups
        independent groups
    b. a one-way ANOVA for   d. a _t_-test for correlated groups
        correlated groups

_____ 26. A researcher is using a repeated measures design with more than two groups. What type of analysis should he use?

    a. a _t_-test for independent groups   c. a one-way ANOVA for
                                       independent groups
    b. a _t_-test for correlated groups   d. a one-way ANOVA for
                                       correlated groups

Matching

a. $k - 1$

b. SS_{within}

c. mean square

d. treatment sum of squares

e. SS_{tot}

f. sum of squares

g. post hoc comparison

h. error sum of squares

i. $N - k$

j. total sum of squares

_____ 1. the "averaged" variability for each score

_____ 2. method for calculating the degrees of freedom for between-groups

_____ 3. symbol for the total sum of squares

_____ 4. the total amount of variability in the data from an experiment

_____ 5. the amount of variability in the data due to sources other than the IV

_____ 6. method for calculating the degrees of freedom for within-groups

_____ 7. the amount of variability in the DV attributable to each source

_____ 8. symbol for the error sum of squares

_____ 9. statistical test for comparing group means after finding a significant F-ratio

_____ 10. the amount of variability in the data due to the IV

True/False

Indicate whether the sentence or statement is true or false.

_____ 1. A t-test is used to statistically analyze multiple group designs.

_____ 2. It is <u>not</u> necessary to use two different types of one-way ANOVAs to analyze multiple-independent-group and multiple-correlated group designs.

_____ 3. To obtain the MS, you must divide the SS by its *df*.

_____ 4. A significant F-ratio does <u>not</u> provide information regarding which of the possible group comparisons is significant.

_____ 5. A post hoc test is used for two-group designs.

_____ 6. Researchers <u>do</u> <u>not</u> need to conduct a post hoc test when they find overall significance in a one-way ANOVA.

_____ 7. The between-groups SS added to the within-group SS should <u>always</u> be equal to the total SS.

_____ 8. The tails of distributions are asymptotic.

_____ 9. The total amount of variability in the data from an experiment is called SS.

_____ 10. A MS represents the variability for each source.

Short Answer

1. Summarize the appropriate statistical analysis procedure for multiple-group designs.

2. Recreate the general conceptual formula for the multiple-group design. Explain why this formula has been altered from the formula previously provided in Chapter 12.

3. Compare and contrast the two-group and multiple-group designs.

4. List the necessary steps for calculating and interpreting a one-way ANOVA for independent samples.

5. List the necessary steps for calculating and interpreting a one-way ANOVA for correlated samples.

6. Describe a source table. What information would you typically find in this table?

7. Define mean squares (MS). Explain what information is needed to calculate a MS. Why are they calculated?

8. Explain what information is needed to calculate an F-ratio. Discuss the logic involved with using the F-ratio.

9. Summarize why there is an additional step involved with calculating a one-way ANOVA for correlated samples.

10. Explain the purpose of a post hoc comparison. In what type of research design is this procedure used?

SPSS Computer Practice Problem 13.1 Using SPSS 11.0 for Windows

Basically when conducting a statistical analysis using SPSS for Windows you will name your variables, enter your data, and finally analyze your data by selecting options from a toolbar. In the following section a step by step procedure of an example problem will be provided to guide you in calculating a one-way ANOVA for independent groups. By following the systematic instructions provided and by referring to the screen figures (SF) when prompted, you should be able to independently conduct a one-way ANOVA for independent groups.

Data

The fictional data for this problem is based on a researcher who studied the influence that a job applicant's first name has on business employers' choice of a job applicant. Essentially, the researcher examined if the business employers' rating of the job applicant would significantly differ based on the first name of the job applicant. The researcher randomly assigned 15 business employers to three equal groups. The researcher used one job resume and changed the first name (no last name was provided) to create three different treatment conditions. The employers in group 1 were given a resume with a traditional first name. The employers in group 2 were given a resume with a non-traditional first name. The employers in group 3 were given a resume that did not have a first name. The researcher then asked the business employers to review and rate the job candidate or resume on a scale of 1 (poor candidate) to 10 (excellent candidate). The raw data for this example is provided in Table 1.

Table 1

Traditional First Name Group 1	Non-traditional First Name Group 2	No First Name Group 3
10	5	4
7	1	6
5	3	9
10	7	3
8	4	3

While working on this example problem consider the following questions:

1. What are the IVs and the DV used in this study?

2. How will you express the null and alternative or experimental hypotheses?

3. Is this a directional or non-directional hypothesis?

4. Are the variances of the two groups equal? How did you reach this conclusion?

5. What are the degrees of freedom for this analysis? How did you reach this conclusion?

6. Are the results of this analysis significant at the pre-set .05 alpha level? Should a post hoc test be conducted?

7. How will you interpret these results in non-statistical language?

Procedure for Completing Example Problem 13.1

1. **Data Entry.** This example will have data for three groups: employer ratings of a resume with a traditional name, employer ratings of a resume with a non-traditional name, and employer ratings of a resume with no name. For the purpose of this example the variable names will be "group" and "jobrate." The variable named "group" will include the three business employer participant groups, however each group will be coded differently so that SPSS recognizes they are separate groups. For example, the participants in the traditional name group will be coded with a "1," the participants in the non-traditional group will be coded with a "2," and the participants in the no name group will be coded with a "3" (see SF-13.1). The variable named "jobrate" will include corresponding information about the employers' rating of the job cadidate. For example, the first participant in screen figure 13.1, row one, column one, will be from the traditional name group ("1") and their corresponding job candidate rating will be a "10." The data for this analysis should resemble screen figure 13.1.

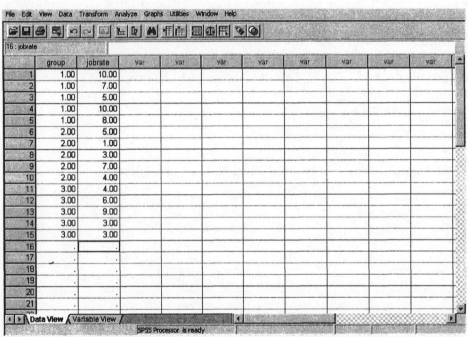

SF-13.1. Entering data to conduct a one-way ANOVA.

Before conducting the one-way ANOVA it may be helpful to label the values you have used in your "group" variable. This will make your output much easier to interpret. Labeling these values can be accomplished by selecting the "Variable View" button on the bottom left corner of the data entry screen (see SF-13.1). Once selected your screen should resemble screen figure 13.2. Next click on the box below the "Values" column for row one. This box should have the word "None" in it before you select it. Once selected a "Value Labels" command box will appear (see SF-13.2). At this point enter a "1" in the "Value" cell and enter the name ("traditional name") of this value in the "Value Label" cell. Then select the "Add" button in the lower left hand corner of the "Value Labels" command box to move this information into the larger white box. If you make a mistake you can use the "Remove" command and start over. Repeat this process for the remaining groups (see SF-13.2). Once you have finished select the "OK" button on the "Value Labels" command box and return to the data entry screen by selecting "Data View" at the bottom left hand corner of the screen (see SF-13.2).

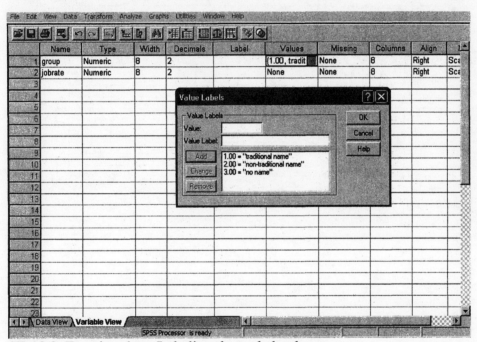

SF-13.2. Entering data: Labeling the coded values.

2. **Conducting the one-way ANOVA.** In order to conduct a statistical analysis using SPSS you will need to select the "Analyze" toolbar option at the top of the screen. A dropdown box will appear providing further data analysis options. Select the "Compare Means" option. After selecting this option SPSS will provide you with further analysis alternatives to the right of the initial dropdown box (see SF-13.3). At this point select the "One-Way ANOVA" option.

SF-13.3. Selecting the one-way ANOVA procedure.

Once you have selected the "One-Way ANOVA" option you should be able to view a command box named "One-Way ANOVA" (see SF-13.4). At this point the variables "jobrate" and "group" will appear in the large white box on the left hand side of the "One-Way ANOVA" command box. Move the "jobrate" variable into the "Dependent List" box by first highlighting (it should appear blue) this variable and then by selecting the arrow in the middle of the screen (see SF-13.4). Next, move the "group" variable into the "Factor" box by selecting the arrow button closest to the "Factor" box. At this point you will need to make a decision regarding which post hoc test you will use for your analysis. Remember that this test will only be necessary if you find a significant *F*-ratio. This can be accomplished by selecting the "Post Hoc" button on the bottom middle portion of the "One-Way ANOVA" command box (see SF-13.4). Once selected you should be able to view a "One-Way ANOVA: Post Hoc Multiple Comparisons" command box (see SF 13.5).

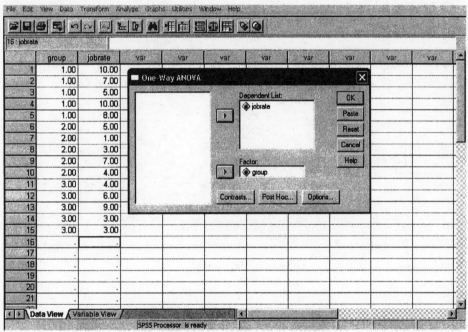

SF-13.4. Conducting the one-way ANOVA procedure.

At this point you should be able to view the "One-Way ANOVA: Post Hoc Multiple Comparisons" command box. Select the "Tukey" option under the "Equal Variances Assumed" heading and close this command box by selecting the "Continue" button on the bottom of the "One-Way ANOVA: Post Hoc Multiple Comparisons" command box. The "Tukey" option is the code for Tukey's honestly significant difference test. Now you will need to make a few more decisions before conducting your analysis. Your screen should once again resemble screen figure 13.4. At this point select the "Options" button on the bottom right side of the "One-Way ANOVA" command box (see SF-13.4). Once selected your screen should resemble screen figure 13.6.

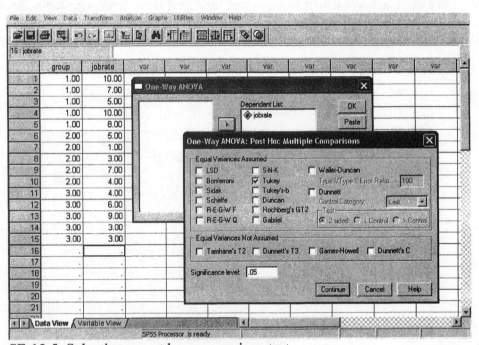

SF-13.5. Selecting a post hoc comparison test.

182

For the purpose of this example you will need to select the "Descriptive" and "Homogeneity of variance test" options under the "Statistics" heading in the "One-Way ANOVA: Options" command box (see SF-13.6). These commands will prompt SPSS to include descriptive statistics and a test of the homogeneity of the group variances in your output. Now select the "Continue" button in the upper right hand corner of the "One-Way ANOVA: Options" command box and SPSS will return you to the "One-Way ANOVA" command box. Finally, select the "OK" button in the upper right hand corner of the "One-Way ANOVA" command box and this will prompt SPSS to run your analysis.

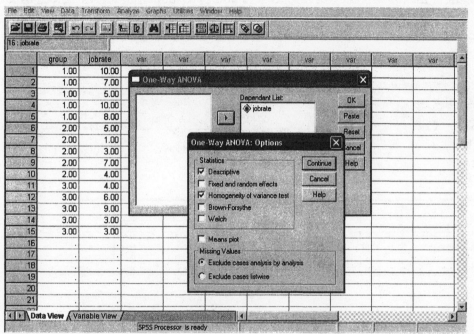

SF-13.6. Selecting options for your one-way ANOVA.

3. **Reviewing the Output for the one-way ANOVA.** Your output should resemble the output in SF-13.7. If it does not you may want to check you raw data for data entry errors. The output in Screen Figure 13.7 provides information regarding the descriptive statistics and Levene's test of the homogeneity of the group variances. The first output labeled "Descriptives," provides the levels of the grouping variables (i.e., traditional name, non-traditional name, and no name), the number of cases, the means and standard deviations, and the standard error for each grouping variable. The second output box provides information about Levene's Test for Equality of Variances. If this test is significant at the .05 alpha level researchers do not assume that the variances of the groups are equal (i.e., they are heterogeneous). Conversely, if this test is not significant, researchers can assume that the variances of the groups are equal (i.e., they are homogeneous).

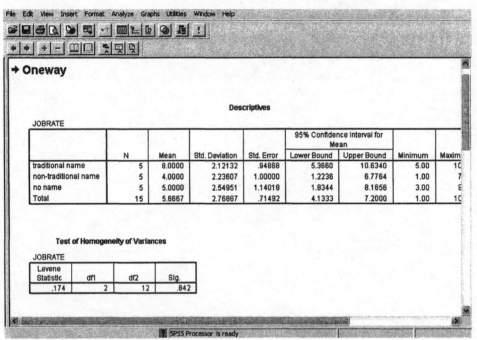

SF-13.7. Output for the one-way ANOVA: Descriptives and Levene's test of homogeneity of variances.

If you scroll the output down you will find the ANOVA table (see SF-13.8). This table lists the types of population variance estimates (i.e., between groups, within groups, and the total), the sum of squares, the degrees of freedom, the mean squares, the *F*-ratio for between groups, and the probability (i.e., "Sig") or the exact chance of obtaining an *F*-ratio this extreme with this particular *F*-distribution.

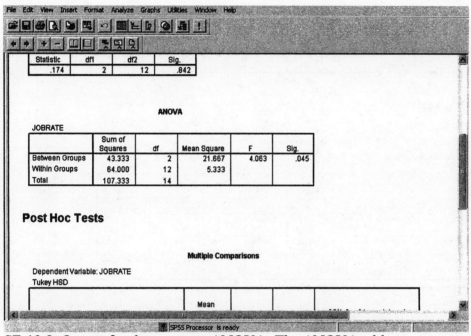

SF-13.8. Output for the one-way ANOVA: The ANOVA table.

If you continue to scroll the output down you will find the results of the Tukey post hoc comparison test (see SF-13.9). Remember this test allows researchers to determine which groups differ significantly from each other once a significant F-ratio has been obtained. Essentially, this test helps researchers to determine where the significance lies in a multiple-group experiment.

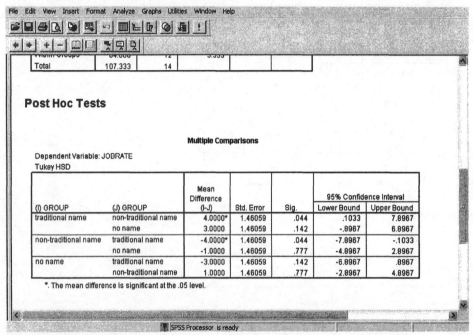

SF-13.9. Output for the one-way ANOVA: The post hoc test table.

4. Interpreting the Results.

<u>What were the IVs and the DV used in this study?</u>

The DV was the employers' rating or measurement of the job candidate or resume on a scale of 1 (poor candidate) to 10 (excellent candidate). The IVs were the three treatment conditions: the resume with a traditional first name, the resume with a non-traditional first name, and the resume that did not have a first name.

<u>How would you express the null and alternative or experimental hypotheses?</u>

$H_0 = M_1 = M_2 = M_3$

$H_a = \underline{not}$ all Ms are equal

<u>Was this a directional or non-directional hypothesis?</u>

This was a non-directional hypothesis because the researcher was only interested in learning if the means between the three treatment conditions would differ: $H_0 = M_1 = M_2 = M_3$.

<u>Were the variances of the two groups homogenous (equal) or heterogeneous (unequal)? How did you reach this conclusion?</u>

In this example the results of the Levene's test for equality of variances was not significant given that the p-value of .842 was much greater than .05 (see SF-13.7). Consequently, the researcher has no reason to doubt the assumption of equal population variances and can assume the group variances are homogeneous.

<u>What were the degrees of freedom for this analysis? How did you reach this conclusion?</u>

The degrees of freedom for all three groups or the "Total" degrees of freedom corresponds to the formula for df_{tot}. The degrees of freedom for the "Between Groups" computer output comparison corresponds to the formula for $df_{between}$. Lastly, the degrees of freedom for the "Within Groups" computer output comparison corresponds to the formula for df_{within} (see SF-13.8).

$$df_{tot} \quad = N - 1 \quad = \quad 15 - 1 \quad = 14$$
$$df_{between} \quad = K - 1 \quad = \quad 3 - 1 \quad = 2$$
$$df_{within} \quad = N - K \quad = \quad 15 - 3 \quad = 12$$

Where:

N = the total number of scores for all groups

K = the number of IV levels or factors

<u>Were the results of this analysis significant at the pre-set .05 alpha level? Should a post hoc test be conducted? If a post hoc test was necessary, what were the results of this test?</u>

Yes, the result was significant [$F(2, 12) = 4.063, p = .045$] given that the p-value was less than the pre-set significance level of .05 (see SF-13.8). Because our F-ratio was significant we can now use the information from our post hoc test to determine which groups differ significantly from each other. If our F-ratio was not significant we would not use the output from the post hoc analysis. Essentially, this test helps researchers to determine where the significance lies in a multiple-group experiment.

Screen figure 13.9 provides the output for our post hoc comparisons. In SF-13.9 we see that the traditional name group was significantly different ($p = .044$) at the .05 alpha level from the non-traditional name group according to the Tukey test. This result means that the business employers rated the job resumes with traditional names ($M = 8.0$) significantly higher than the job resumes with non-traditional names ($M = 4.0$). No other groups differed significantly from each other, meaning that there were no significant differences in the business employers' ratings of the resumes of the job candidates for the following comparisons:

1. traditional name and no name provided ($p = .142$).
2. non-traditional name and no name provided ($p = .777$).

How would you interpret these results in non-statistical language?

Overall, the business employers' choice of a job applicant, based solely on the ratings of the applicant resumes, significantly differed depending on the first name of the job applicant. The business employers' choice of a job applicant was significantly increased when the applicant had a traditional first name compared with an applicant who's name was not provided or with an applicant who had a non-traditional first name.

Answers for the multiple choice, matching, and true/false items

Multiple Choice		Matching		True and False	
1.	D	1.	C	1.	F
2.	C	2.	A	2.	F
3.	B	3.	E	3.	T
4.	C	4.	J	4.	T
5.	C	5.	H	5.	F
6.	A	6.	I	6.	F
7.	C	7.	F	7.	T
8.	B	8.	B	8.	T
9.	D	9.	G	9.	F
10.	A	10.	D	10.	T
11.	A				
12.	A				
13.	C				
14.	B				
15.	C				
16.	D				
17.	D				
18.	A				
19.	C				
20.	D				
21.	B				
22.	D				
23.	E				
24.	D				
25.	A				
26.	D				

Chapter 14

Designing and Conducting Experiments with Multiple Independent Variables

Learning Objectives

By the end of this chapter you should be able to:

1.   Summarize the necessary components of a factorial design.

2.   Contrast and compare factorial designs with two-group and multiple-group designs.

3.   Explain why factorial designs are critical to experimental research.

4.   Discuss why researchers must assume more responsibilities when using factorial designs.

5.   Describe the procedures for constructing shorthand notations for factorial designs.

6.   Define and distinguish between a main effect and an interaction.

7.   Summarize the difficulties that researchers encounter when using methods for creating correlated participant groups in factorial designs.

8.   Explain why 2 x 2 factorial designs are appropriate for totally repeated measures designs.

9.   Discuss the benefits and limitations of using factorial designs with mixed assignments of participant groups.

10.   Summarize the implications associated with interpreting information from an experiment that uses both manipulated and measured IVs.

11.   List the possible interaction effects of an experiment with four IVs.

12.   Discuss the potential problems associated with using a repeated measures factor in your factorial experiment.

13.   Define and draw a diagram depicting an interaction.

14.   Explain how the KISS principle relates to experimental research designs.

Expanded Outline

I.   Experimental Design: Doubling the Basic Building Block
                        Factors or IVs
                The Factorial Design

                        How Many IVs?
                                Factorial Design
                        How Many Groups or Levels?

                                Main Effects
                                Interactions

                        Assigning Participants to Groups
                                Independent Groups
                                Correlated Groups
                                Mixed Assignment
                        Random Assignment to Groups
                                Between-Subjects Factorial Designs
                        Nonrandom Assignment to Groups
                                Completely Within-Groups
                                        Matched Pairs or Sets
                                        Repeated Measures

                                        Natural Pairs or Sets

                        Mixed Assignment to Groups
                                Mixed Factorial Designs

II.   Review Summary

III.   Check Your Progress

IV.   Choosing an Experimental Design
                Comparing the Factorial Design to Two-Group and Multiple-Group Designs

                Choosing a Factorial Design
                        Experimental Questions
                        Control Issues
                        Practical Considerations
                                Principle of Parsimony
                                KISS

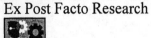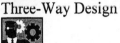
Practice Exam

Multiple Choice

Identify the letter of the choice that best completes the statement or answers the question.

_____ 1.   Factorial designs are important because they allow researchers to simultaneously examine

    a. descriptive data.   c. combinations of IVs.
    b. one IV.   d. none of the above

_____ 2.   The simplest possible factorial design is expressed in shorthand notation as:

    a. 1 x 1   c. 3 x 3
    b. 2 x 2   d. 4 x 4

_____ 3.   Researchers named the factorial design based on multiple

    a. nuisance variables.   c. DVs.
    b. IVs.   d. extraneous variables.

____ 4. In factorial designs the _____ of numbers describes the amount of IVs; whereas the _____ of each number describes the amount of levels for each IV.

    a. number, value   c. value, square root
    b. value, number   d. number, standard deviation

____ 5. Molly is conducting an experiment with two IVs. The first IV has two levels and the second IV has four levels. How would this design be expressed in shorthand notation?

    a. 1 x 4   c. 2 x 4
    b. 4 x 1   d. 2 x 2

____ 6. A researcher is studying the effect of gender and major (math, English, and history) on freshmen GPA. How should this design be expressed in shorthand notation?

    a. 1 x 2   c. 2 x 3
    b. 2 x 1   d. 3 x 1

____ 7. The fewest treatment conditions possible in a factorial design is

    a. 1   c. 3
    b. 2   d. 4

____ 8. The _____ refers to the sole effect of one IV in a factorial design; whereas the joint, simultaneous effect of more than one IV on the DV is known as the _____.

    a. main effect, interaction   c. interaction, Principle of Parsimony
    b. interaction, main effect   d. main effect, Principle of Parsimony

____ 9. The primary advantage of using a factorial design instead of a two-group design is the

    a. main effect.   c. ability to use more IVs.
    b. interaction effect.   d. design is more parsimonious.

____ 10. An interaction can be described as

    a. the effects of one IV change as the levels of the other IV change.   c. lines on a graph that are not parallel.
    b. the effects of one IV depend on the particular level of another IV.   d. all of the above

_____ 11. Sam is using a factorial ANOVA to examine the effects of two IVs. Which outcome should he initially evaluate for significance?

    a. the main effect of the first IV   c. the interaction of both IVs
    b. the main effect of the second IV   d. both a and b

_____ 12. In a factorial design participants can be grouped by

    a. random assignment.   c. both random and correlated assignment.
    b. correlated assignment.   d. all of the above

_____ 13. Factorial designs in which both IVs involve random assignment may be called

    a. completely within-group designs.   d. both a and b
    b. between-subjects factorial designs.   e. both b and c
    c. completely randomized designs.

_____ 14. A researcher wants to conduct a study using a 3 x 3 factorial design with matched pairs or sets. How many matched participant groups will be required to conduct this study?

    a. 3   c. 9
    b. 6   d. 12

_____ 15. Oliver is conducting a factorial study with two IVs. He has randomly assigned the participants to the first IV and used correlated assignment to group participants for the second IV. What method of assignment is he using for his study?

    a. independent   c. correlated
    b. mixed   d. non-random

_____ 16. Which method for creating correlated groups is often the most difficult to use with human participants; consequently researchers rarely use this approach other than for littermates.

    a. natural pairs or sets   c. random assignment
    b. repeated measures   d. matched pairs or sets

_____ 17. In a two-IV mixed assignment factorial design, the use of _____ is probably more likely than other types of non-random assignment.

    a. repeated measures   c. matched pairs or sets
    b. natural pairs or sets   d. random assignment

18. Marshall is reviewing the graphical results of his factorial experiment and notices that the lines on the graph are <u>not</u> parallel (i.e., they intersect). What should Marshall conclude from this information?

    a. the main effects were significant   c. the interaction was significant
    b. the main effects were not significant d. the interaction was not significant

19. Researchers typically assume that random assignment of participants to groups will adequately equate the groups if there is approximately _____ participants per group.

    a. 10   c. 30
    b. 20   d. 40

20. The simplest possible factorial design with three IVs is referred to as a _____ design.

    a. one-way   c. three-way
    b. two-way   d. four-way

21. A researcher has conducted a study using three IVs (A, B, and C). How many interactions should this researcher expect to find?

    a. 2   c. 4
    b. 3   d. 5

22. Annie has conducted a study using four IVs (A, B, C, and D) resulting in a 2 x 4 x 3 x 3 experimental design. How many interactions should she expect to find?

    a. 4   c. 8
    b. 6   d. 11

Matching

a. factorial design   f. graphical depiction of a <u>non</u> interaction

b. 2 x 2   g. interaction

c. 2 x 2 x 2   h. main effect

d. ex post facto   i. graphical depiction of an interaction

e. 3 x 3   j. factors

_____ 1. lines on a graph that are parallel

_____ 2. refers to the sole effect of one IV in a factorial design

_____ 3. shorthand notation for an experimental design with three IVs

_____ 4. an experimental design with more than one IV

_____ 5. the joint, simultaneous effect of more than one IV on the DV

_____ 6. a term that is synonymous with IVs

_____ 7. a term for using measured variables in a factorial design

_____ 8. lines on a graph that are <u>not</u> parallel

_____ 9. shorthand notation for the simplest factorial design

_____ 10. shorthand notation for a factorial design with two IVs each having three levels

True/False

Indicate whether the sentence or statement is true or false.

____ 1. Factorial designs allow for the simultaneous examination of a combination of IVs.

____ 2. There is <u>no</u> limit to the number of IVs that can be used in an experiment.

____ 3. The smallest amount of treatment conditions possible in a factorial design is 3.

____ 4. The simplest way to identify an interaction is to look for lines on a graph that are <u>not</u> parallel.

____ 5. A 3 x 3 design would most likely be a candidate for a totally repeated measures study.

____ 6. The simplest factorial design is notationally expressed as 1 x 2.

____ 7. The most difficult method for creating correlated groups with human participants is natural pairs or sets.

____ 8. Mixed assignment designs involve a combination of random and nonrandom participant assignment procedures.

____ 9. There is <u>no</u> limit to the number of levels for any IV in a factorial design.

____ 10. In factorial designs the number of IV levels <u>cannot</u> be unequal.

____ 11. The true advantage of factorial designs is that they measure interactions.

Short Answer

1. Summarize the necessary components of a factorial design.

2. Contrast and compare factorial designs with two-group and multiple-group designs.

3. Explain why factorial designs are critical to experimental psychology and why they more closely resemble the real world.

4. Discuss why researchers must assume more responsibilities when using factorial designs.

5. Explain how researchers construct shorthand notations for factorial designs. Create a research study that would require a 3 x 3 design. How many treatment combinations would be involved in a 3 x 3 design? Draw a diagram depicting this design.

6. Define and distinguish between a main effect and an interaction. Which of these do researchers typically examine first?

7. Summarize the difficulties that researchers encounter when using methods for creating correlated participant groups in factorial designs.

8. Explain why 2 x 2 factorial designs are appropriate for totally repeated measures designs.

9. Discuss the benefits and limitations of using factorial designs with mixed assignments of the participant groups. Try to construct an example of a factorial design that uses mixed assignment of participant groups.

10. Summarize the implications associated with interpreting information from an experiment that uses both manipulated and measured IVs.

11. Draw a diagram depicting an experiment with four IVs. List the possible main effects and interactions.

12. Discuss the potential problems associated with using a repeated measures factor in your factorial experiment. How can researchers address these problems?

13. Define and draw a diagram depicting an interaction.

14. Explain how the KISS principle relates to experimental research designs.

Answers for the multiple choice, matching, and true/false items

Multiple Choice		Matching		True and False	
1.	C	1.	F	1.	T
2.	B	2.	H	2.	T
3.	B	3.	C	3.	F
4.	A	4.	A	4.	T
5.	C	5.	G	5.	F
6.	C	6.	J	6.	F
7.	B	7.	D	7.	T
8.	A	8.	I	8.	T
9.	B	9.	B	9.	T
10.	D	10.	E	10.	F
11.	C			11.	T
12.	D				
13.	E				
14.	C				
15.	B				
16.	A				
17.	A				
18.	C				
19.	A				
20.	C				
21.	C				
22.	D				

Chapter 15

Analyzing and Interpreting Experiments with Multiple Independent Variables

Learning Objectives

By the end of this chapter you should be able to:

1. Summarize the rational for using factorial designs.

2. Explain what information can be obtained from a two-way ANOVA that cannot be obtained from two one-way ANOVA designs using the same variable.

3. Construct and describe the conceptual formulas for two-IV factorial designs.

4. Explain the relation between synergistic effects and experimental research.

5. List and describe the different sources for between-groups variability.

6. Discuss why it is wise to limit a multi-factor experiment to two or three factors.

7. Explain why post hoc tests may not be necessary in two-way ANOVAs.

8. Summarize how a source table for a two-way ANOVA differs from a source table for a one-way ANOVA.

9. Define the terms main effect and interaction and explain why a significant interaction renders its associated main effects uninterpretable.

10. Discuss why researchers do not test for the participant effects when using a two-way repeated measures ANOVA for correlated groups.

11. Describe the necessary components of a two-way ANOVA for mixed groups.

12. Explain the benefits and limitations of a two-way ANOVA for mixed groups.

By the end of this chapter you should learn when and how to use the following formulas:

I.   The steps for calculating a two-way ANOVA for independent groups

1. First calculate the sums of all the scores for each group.
    $$\left(i.e., \quad \Sigma X_1 + \Sigma X_2 + ...\Sigma X_n ... = \Sigma X_{tot} \right)$$

2. Then square each of the participant scores and sum these squared values.
    $$\left(i.e., \quad X_1^2 + X_2^2 + X_n^2 ... = \Sigma X_{tot}^2 \right)$$

3. Next calculate the total sum of squares (SS_{tot}) using formula 1.

    $$SS_{tot} = \Sigma X_{tot}^2 - \left[\frac{\left(\Sigma X_{tot} \right)^2}{N} \right]$$

    (1)

4. Now calculate the main effect sums of squares or treatment sum of squares (i.e., SS_A and SS_B). Given that this is a two-way ANOVA or that we have two IVs, we must compute the variability for each group separately. The order in which you compute the main effect sums of squares is not important. This can be accomplished by using formulas 2 and 3.

    $$SS_A = \Sigma \left[\frac{(sum\ of\ scores\ in\ each\ column)^2}{n\ of\ scores\ in\ the\ column} \right] - \left[\frac{(\Sigma X_{tot})^2}{N} \right]$$

    (2)

    $$SS_B = \Sigma \left[\frac{(sum\ of\ scores\ in\ each\ column)^2}{n\ of\ scores\ in\ the\ column} \right] - \left[\frac{(\Sigma X_{tot})^2}{N} \right]$$

    (3)

5. Next, calculate the interaction sum of squares (i.e., SS_{AxB}) by beginning with the cell mean (ex., $M_{A_1B_1}$), subtracting the two overall means for its column and row, and then adding the overall mean (M_{tot}). For each case, you will square the total, sum those values, and then multiply this value by the number of participants per cell (i.e., n). Before calculating the interaction sum of squares it may be helpful to visualize the comparison matrix (see Figure 1) and formula 4. Now you are ready to calculate the interaction sum of squares using formula 4.

Figure 1

	Factor B (first IV)	
	Level B₁	**Level B₂**
Level A₁	**A₁ B₁**	**A₁ B₂**
Level A₂	**A₂ B₁**	**A₂ B₂**

(with "Factor A (second IV)" as a vertical label on the left)

$$SS_{AxB} = n \begin{bmatrix} (M_{A_1 B_1} - M_{column B_1} - M_{row A_1} + M_{tot})^2 + \\ (M_{A_1 B_2} - M_{column B_2} - M_{row A_1} + M_{tot})^2 + \\ (M_{A_2 B_1} - M_{column B_1} - M_{row A_2} + M_{tot})^2 + \\ (M_{A_2 B_2} - M_{column B_2} - M_{row A_2} + M_{tot})^2 \end{bmatrix}$$

(4)

6. Next, calculate the <u>error sum of squares</u> (i.e., SS_{error}) using the formula for the total sum of squares (formula 1) and applying it to each of the four cells (as shown in Figure 1) in the two-way ANOVA as shown in formulas 5-8.

$$cell_{A_1 B_1} = \Sigma X^2_{A_1 B_1} - \left[\frac{\left(\Sigma X_{A_1 B_1} \right)^2}{n_{A_1 B_1}} \right]$$

(5)

$$cell_{A_1 B_2} = \Sigma X^2_{A_1 B_2} - \left[\frac{\left(\Sigma X_{A_1 B_2} \right)^2}{n_{A_1 B_2}} \right]$$

(6)

$$cell_{A_2B_1} = \Sigma X^2_{A_2B_1} - \left[\frac{\left(\Sigma X_{A_2B_1} \right)^2}{n_{A_2B_1}} \right]$$

(7)

$$cell_{A_2B_2} = \Sigma X^2_{A_2B_2} - \left[\frac{\left(\Sigma X_{A_2B_2} \right)^2}{n_{A_2B_2}} \right]$$

(8)

7. Now find the <u>mean square</u> for each source of variation in order to make them directly comparable. However, in order to calculate the mean squares, you first need to determine the <u>degrees of freedom</u> for each source of variation. This can be accomplished by using the following formulas:

<u>Total degrees of freedom</u>

$$df_{tot} = N - 1$$

<u>Main effect degrees of freedom</u>

$$df_A = k_A - 1$$

$$df_B = k_B - 1$$

<u>Interaction degrees of freedom</u>

$$df_{AxB} = (df_A)(df_B)$$

<u>Error degrees of freedom</u>

$$df_{error} = N - (k_A)(k_B)$$

<u>Where:</u>

N = the total number of participants in the study
k = the total number of levels or groups in the study
k_A = the number of levels in factor A
k_B = the number of levels in factor B

8. Next, calculate the <u>mean squares</u> for each source of variation by dividing each sum of squares by its respective degrees of freedom. This can be accomplished by using formulas 9-12.

<u>Mean squares for the main effects</u>

$$MS_A = \frac{SS_A}{df_A}$$

(9)

$$MS_B = \frac{SS_B}{df_B}$$

(10)

<u>Mean square for the interaction</u>

$$MS_{AxB} = \frac{SS_{AxB}}{df_{AxB}}$$

(11)

<u>Mean square for the error</u>

$$MS_{error} = \frac{SS_{error}}{df_{error}}$$

(12)

9. Then, calculate the <u>F-obtained ratios</u> by dividing the mean square for each source of variation by the error or within mean square by using formulas 13-15.

<u>F-ratios for the main effects</u>

$$F_A = \frac{MS_A}{MS_{error}}$$

(13)

$$F_B = \frac{MS_B}{MS_{error}}$$

(14)

<u>F-ratio for the interaction</u>

$$F_{AxB} = \frac{MS_{AxB}}{MS_{error}}$$

(15)

10. Lastly, calculate the <u>effect size</u> (η^2) as a measure of the proportion of the variance accounted for by an effect. Assuming that your interaction was significant, use formula 16 to calculate the effect size for the interaction.

$$\eta^2_{AxB} = \frac{(F_{AxB})\,(df_{AxB})}{(F_{AxB})\,(df_{AxB}) + df_{error}}$$

(16)

Expanded Outline

I. Before Your Statistical Analysis
    Analyzing Factorial Designs

    Planning Your Experiment

    Rationale of Factorial ANOVA
        Treatment Variability
        Error Variability
    Understanding Interactions
        Synergistic Effects

II. Two-way ANOVA for Independent Groups
    Calculating ANOVA by Hand

        Sum of Squares
        Total Sum of Squares
        Main Effect Sums of Squares
        Interaction Sum of Squares
        Error Sum of Squares
        Mean Squares
            Degrees of Freedom (*df*)
        F-ratios

        Post Hoc Comparisons
        Source Table

        Effect Size

Post Hoc Tests
Source Table
Effect Size
Computer Analysis of Two-way ANOVA for Mixed Samples
Interpreting Computer Statistical Output

Translating Statistics into Words

A Final Note

Practice Exam

Multiple Choice

Identify the letter of the choice that best completes the statement or answers the question.

_____   1.   Researchers analyze factorial designs with

      a. ANOVA.   c. *t*-tests.
      b. regression.   d. correlations.

_____   2.   Which label describes how researchers assign participants to groups?

      a. independent groups   d. totally between-subjects
      b. completely between-subjects   e. all of the above
      c. completely between-groups

_____ 3. Research designs using matching or repeated measures may be called

    a. randomized block.   d. totally within-groups.
    b. completely within-subjects.   e. all of the above
    c. completely within-groups.

_____ 4. Researchers typically ignore the main effects of the IVs when they find a(n) _____ interaction.

    a. non-significant   c. biased
    b. significant   d. inconclusive

_____ 5. Which model is <u>not</u> included in the two-way ANOVA?

    a. fixed effects   c. mixed effects
    b. random effects   d. expected effects

_____ 6. A significant interaction can be identified when the

    a. lines on a graph intersect.   d. both a and c
    b. lines on a graph are parallel. e. both b and c
    c. effect of one IV depends on the
       specific level of the other IV.

_____ 7. Which of the following is <u>not</u> a variability component for a between-groups two-way ANOVA?

    a. the first IV   c. the second IV
    b. the first DV   d. the interaction of the two IVs

_____ 8. The averaged variability for each source in a two-way ANOVA is referred to as

    a. mean squares.   c. degrees of freedom.
    b. sum of squares.   d. interaction.

_____ 9. In a two-way ANOVA researchers calculate F-obtained ratios for the

    a. main effects.   d. both a and b
    b. interaction effects.   e. both b and c
    c. error effects.

____ 10. When conducting a two-way ANOVA, which sum of squares is computed, but not displayed in the summary table, and not used to obtain a mean square or F-obtained ratio?

a. the main effect for the first IV   c. the interaction of both IVs
b. the main effect for the second IV   d. the between-groups effect

____ 11. An advantage of the two-factor design is that

a. it is possible to test for an interaction. c. greater generalizability is possible.
b. fewer participants are needed.   d. all of the above

____ 12. In a two-factor experiment, the effect of each independent IV is called a(n)

a. main effect.   c. factorial effect.
b. interaction.   d. dependent effect.

____ 13. Which of the following is not considered to be an advantage of the factorial design over the one-way ANOVA?

a. a constant alpha level   c. control over additional variables
b. efficiency   d. the ability to examine interactions

____ 14. How many treatment groups are present when using a 2 x 3 factorial design?

a. 5   c. 8
b. 6   d. 9

____ 15. Which procedure allows researchers to identify means that are significantly different after obtaining a significant F-obtained value?

a. coefficient of determination   c. z-scores
b. effect sizes   d. post hoc tests

____ 16. Plotting cell means is typically done to examine

a. main effects.   c. interaction.
b. simple effects.   d. association effects.

____ 17. In a two-way ANOVA, post hoc tests may not be necessary because the

a. interaction is significant and   c. multiple-level IV is not significant.
   supersedes the main effects.
b. experimental design makes   d. all of the above
   them unnecessary.

_____ 18. Typically the effect size for a two-way ANOVA for _____ groups will be larger than it would be for a two-way ANOVA for _____ groups.

    a. independent, random   c. correlated, independent
    b. independent, correlated   d. none of the above

_____ 19. In a correlated samples design the _F_-obtained ratios are _____ and the probabilities are _____ for both the IVs and the interaction.

    a. smaller, smaller   c. larger, smaller
    b. smaller, larger   d. larger, larger

_____ 20. A two-way ANOVA requiring IVs with independent groups for one IV and correlated groups for the second IV is called

    a. mixed groups.   c. correlated groups.
    b. equal groups.   d. randomized groups.

_____ 21. A series of research experiments with a related topic or question is called

    a. categorical research.   c. marginal research.
    b. programmatic research.   d. post hoc research.

_____ 22. The analysis technique for experiments with two IVs that are both between-subjects factors is referred to as a two-way ANOVA for

    a. independent groups.   c. mixed groups.
    b. correlated groups.   d. post hoc groups.

Matching

a. η^2

b. post hoc

c. synergistic effects

d. interaction sum of squares

e. sum of squares

f. programmatic research

g. correlated groups

h. total sum of squares

i. independent groups

j. *F*-obtained

k. treatment sum of squares

l. mixed group design

____ 1. used to calculate an effect size

____ 2. the amount of variability due to an IV

____ 3. a two-way ANOVA with independent and correlated groups

____ 4. series of experiments that involve a related topic

____ 5. a two-way ANOVA comprised of only within-subject factors

____ 6. the amount of variability due to an interaction

____ 7. a two-way ANOVA with only between-subject factors

____ 8. the amount of variability in the DV attributable to each source

____ 9. the total amount of variability in DV scores

____ 10. a ratio calculated by dividing a treatment mean square by an error mean square

____ 11. a consequence occurring from the combination of two or more conditions

____ 12. a test used after finding a significant *F*-obtained value

True/False

Indicate whether the sentence or statement is true or false.

_____ 1. Factorial designs are analyzed using analysis of variance.

_____ 2. Designs using a mixture of between and within assignment procedures may be referred to as completely randomized.

_____ 3. Researchers virtually ignore the main effects of an IV when they obtain a significant interaction.

_____ 4. Researchers interpret interactions by graphing the IV on the y-axis and the DV on the x-axis.

_____ 5. An interaction is present when the effect of one IV depends on the specific level of the other IV.

_____ 6. Main effect sum of squares are analogous to treatment sum of squares.

_____ 7. Independently F-obtained ratios allow researchers to determine which of several means are significantly different.

_____ 8. In a two-way ANOVA post hoc tests may <u>not</u> be necessary.

_____ 9. The two-way ANOVA for correlated groups analysis technique uses two IVs that are both between-subjects factors.

_____ 10. The two-way ANOVA for mixed groups analysis technique uses two IVs that are both within-subjects factors.

_____ 11. Programmatic research involves a series of research experiments that deal with a related topic or question.

Short Answer

1. Summarize the rationale for using factorial designs. Discuss why the sources of treatment variability increase with this design.

2. Explain what information can be obtained from a two-way ANOVA that cannot be obtained from two one-way ANOVA designs using the same variable.

3. Construct and describe the conceptual formulas for two-IV factorial designs.

4. Explain the relation between synergistic effects and experimental research.

5. List and describe the different sources for between-groups variability.

6. Discuss why it is wise to limit a multi-factor experiment to two or three factors.

7. Explain the reasons why post hoc tests may not be necessary in using two-way ANOVAs.

8. Summarize how a source table for a two-way ANOVA differs from a source table for a one-way ANOVA.

9. Define the terms main effect and interaction. Explain why a significant interaction renders its associated main effects uninterpretable.

10. Discuss why researchers do not test for the participant effects when using a two-way repeated measures ANOVA for correlated groups.

11. Describe the necessary components of a two-way ANOVA for mixed groups. Explain the benefits and limitations of this design. How does this design differ from the two-way ANOVAs for independent and correlated groups?

SPSS Computer Practice Problem 15.1 Using SPSS 11.0 for Windows

Basically when conducting a statistical analysis using SPSS for Windows you will name your variables, enter your data, and finally analyze your data by selecting options from a toolbar. In the following section a step by step procedure of an example problem will be provided to guide you in calculating a two-way ANOVA for independent groups. By following the systematic instructions provided and by referring to the screen figures (SF) when prompted, you should be able to independently conduct a two-way ANOVA for independent groups.

Data

The fictional data for this problem is based on a researcher who studied the influence that meditation and sex would have on air traffic controllers' perceived level of stress. Essentially, the researcher examined if the perceived level of stress for the air traffic controllers would differ based on their sex and whether or not they were using meditation. The researcher randomly assigned 10 men and 10 women to either the meditation or the non-meditation participant groups (see Table 1). The men and women in the meditation group used meditation practices immediately before and after their eight hour shift. The men and women in the non-meditation group <u>did</u> <u>not</u> use any meditation practices. After one month the researcher asked the air traffic controllers to rate their perceived level of stress on a scale of 1 (minimal stress) to 10 (high stress). The raw data for this example are provided in Table 2.

Table 1

Relaxation Technique	Sex of Air Traffic Controller	
	Males (1)	Females (2)
Meditation	N = 5	N = 5
No Meditation	N = 5	N = 5

Table 2

Sex of the Participants	Meditation Group	NonMeditation group
1	1	5
1	1	6
1	1	2
1	1	2
1	1	5
1	0	10
1	0	10
1	0	9
1	0	10
1	0	10
2	1	6
2	1	7
2	1	5
2	1	3
2	1	4
2	0	5
2	0	6
2	0	5
2	0	7
2	0	6

While working on this example problem consider the following questions:

1. What are the IVs and the DV used in this study?

2. What are the degrees of freedom for this analysis?

3. Will the results of this analysis be significant at the pre-set .05 alpha level? Will a post hoc test be necessary for this analysis? Why or why not?

4. How will you write the statistical results of this experiment?

5. How will you interpret these results in non-statistical language?

Procedures for Completing Example Problem 15.1

1. **Data Entry.** This example will be comprised of data for four participant groups: male air traffic controllers who used meditation practices ($N = 5$), male air traffic controllers who did not use meditation practices ($N = 5$), female air traffic controllers who used meditation practices ($N = 5$), and female air traffic controllers who did not use meditation practices ($N = 5$). For the purpose of this example the variable names will be "sex" ("1" = male and "2" = female), "meditate" ("0" = meditation not used and "1" = meditation used) and "stress" (scored from "1" = minimal stress to "10" = maximum stress). The variable named "sex" will reflect the sex of the participant. The variable named "meditate" will reflect those participants who used or did not use meditation practices. Lastly, the variable named "stress" will reflect the perceived stress level scores for the participants. The participants will be coded differently so that SPSS recognizes they are separate groups. For example, the first participant in screen figure 15.1, row one, column one, is a male ("1") who used meditation ("1") and received a perceived stress score of "5". The data for this analysis should resemble screen figure 15.1.

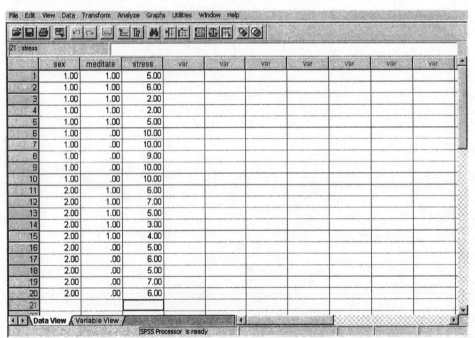

SF-15.1. Entering data to conduct a two-way ANOVA.

216

Before conducting the two-way ANOVA it may be helpful to label the values you have used to describe the levels of your variables. This will make your output much easier to interpret. Labeling these values can be accomplished by selecting the "Variable View" button on the bottom left corner of the data entry screen (see SF-15.1). Once selected your screen should resemble screen figure 15.2. Next click on the box below the "Values" column for row one. This box should have the word "None" in it before it is selected. Once selected a "Value Labels" command box will appear (see SF-15.2). At this point label the values of your sex variable by entering a "1" in the "Value" cell and enter the name ("male") of this value in the "Value Label" cell. Then select the "Add" button in the lower left hand corner of the "Value Labels" command box to move this information into the larger white box. If you make a mistake you can use the "Remove" command and start over. Repeat this process for the "meditate" variable (see SF-15.2). It is not necessary to code the values for the perceived stress score variable ("stress"). Once you have finished select the "OK" button on the "Value Labels" command box and return to the data entry screen by selecting "Data View" at the bottom left hand corner of the screen (see SF-15.2).

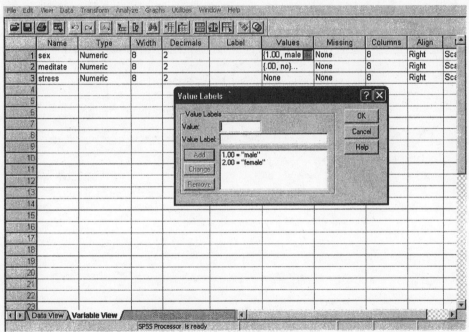

SF-15.2. Entering data: Labeling the coded values.

2. **Conducting the two-way ANOVA.** In order to conduct a statistical analysis using SPSS you will need to select the "Analyze" toolbar option at the top of the screen. A dropdown box will appear providing further data analysis options. Select the "General Linear Model" option. After selecting this option SPSS will provide you with further analysis alternatives to the right of the initial dropdown box (see SF-15.3). At this point select the "Univariate" option.

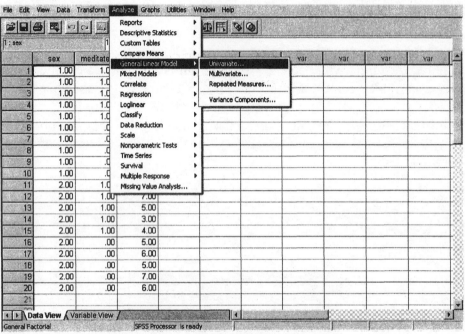

SF-15.3. Selecting the two-way ANOVA procedure.

Once you have selected the "Univariate" option you should be able to view a command box named "Univariate" (see SF-15.4). At this point the variables "sex," "meditate," and "stress" will appear in the large white box on the left hand side of the "Univariate" command box. Move the "stress" variable into the "Dependent Variable" box by first highlighting (it should appear blue) this variable and then by selecting the arrow closest to the "Dependent Variable" box (see SF-15.4). Next, move the variables "sex" and "meditate" into the "Fixed Factor(s)" box by selecting the arrow button closest to the "Fixed Factor(s)" box. Now you will need to make a few more decisions regarding data plots, post hoc tests, and options for your analysis output. Choosing a data plot for your analysis can be accomplished by selecting the "Plots..." button in the upper right hand corner of the "Univariate" command box. Once selected your screen should resemble screen figure 15.5.

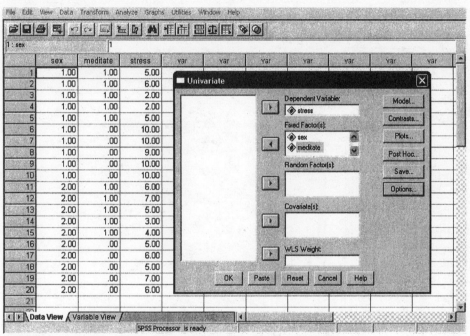

SF-15.4. Conducting the two-way ANOVA procedure.

Once you have selected the "Plots…" option you should be able to view the "Univariate: Profile Plots" command box. Move the "sex" variable to the "Horizontal Axis" box and the "meditate" variable to the "Separate Lines" box. Then select the "Add" button at the bottom left hand corner of screen figure 15.5. At this point you should see "meditate*sex." This selected comparison will prompt SPSS to generate a plot of your data. Now you can close this screen by selecting the "Continue" command at the top right hand corner of the "Univariate: Profile Plots" command box. Next, if appropriate, you can select a post hoc test (ex., Tukey) for your analysis by selecting the "Post Hoc…" command in the "Univariate" command box (SF-15.4). Once selected a "Univariate: Post Hoc Multiple Comparisons for Observed Means" command box will appear. Now, move your variables (i.e., "sex" and "meditate") into the "Post Hoc Tests for:" box and select "Continue." Your screen should once again resemble screen figure 15.4. Lastly, select the "Options" button on the middle right side of the "Univariate" command box (see SF-15.4). Once selected your screen should resemble screen figure 15.6.

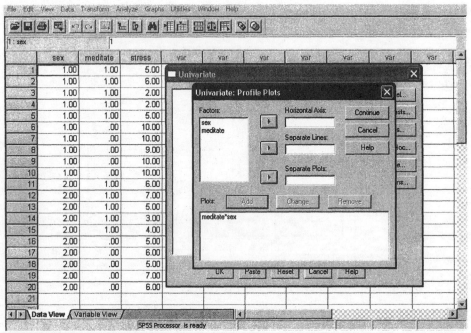

SF-15.5. Selecting the data profile plot for your analysis.

For the purpose of this example select the "Descriptive statistics" and "Estimates of effect size" options under the "Display" heading in the "Univariate: Options" command box (see SF-15.6). These commands will prompt SPSS to include descriptive statistics and an effect size estimate in your output. Now select the "Continue" button in the lower middle portion of the "Univariate: Options" command box and SPSS will return you to the "Univariate" command box (see SF-15.4). Finally, select the "OK" button in the lower left corner of the "Univariate" command box and this will prompt SPSS to run your analysis.

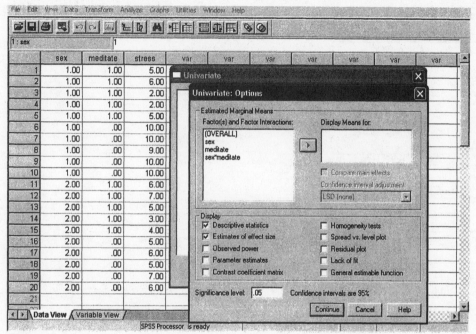

SF-15.6. Selecting output options for your two-way ANOVA.

3. **Reviewing the Output for the two-way ANOVA.** Your output should resemble the output in SF-15.7. If it does not you may want to check your raw data for data entry errors. The output in Screen Figure 15.7 provides information regarding the descriptive statistics. More specifically, it provides the levels of the grouped variables (i.e., "sex" and "meditate"), the corresponding means and standard deviations, and the number of cases for the grouped variables ($N = 10$).

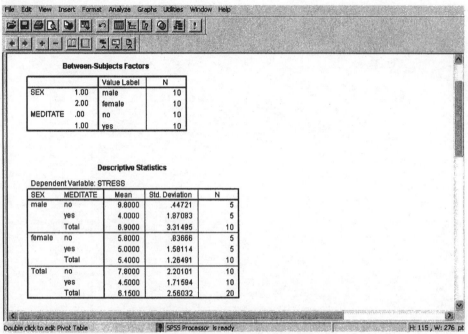

SF-15.7. Output for the two-way ANOVA: Descriptive statistics.

If you scroll the output down you will find the "Tests of Between-Subjects Effects" table for your two-way ANOVA (see SF-15.8). This table lists the types of population variance estimates (e.g., between groups, within groups, and the total), the sum of squares, the degrees of freedom, the mean squares, the corresponding F-ratios, the probability (i.e., "Sig") or the exact chance of obtaining an F-ratio this extreme with this particular F-distribution, and the effect sizes (i.e., Partial Eta Squared). Essentially, this table provides information about the main effects of "SEX" and "MEDITATE" as well as the interaction effect of SEX*MEDITATE.

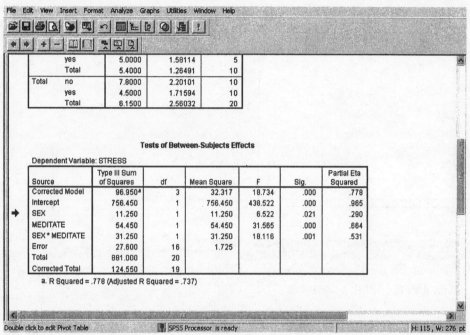

SF-15.8. Output for the two-way ANOVA: The tests of between-subjects effects table.

If you continue to scroll the output down you will find a graph that plots the mean perceived stress scores for the participants by the "sex" and "meditate" variables (see SF-15.9). The broken line (i.e., - -) depicts the mean perceived stress scores for the men; whereas the solid line depicts the mean perceived stress scores for the women.

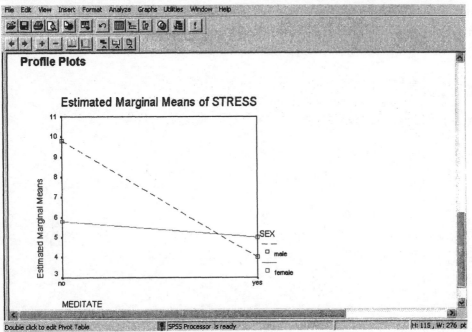

SF-15.9. Output for the two-way ANOVA: Plotted mean stress scores by the sex and meditate variables.

4. Interpreting the Results.

What were the IVs and the DV used in this study?

This example used a 2 (sex) x 2 (meditate) ANOVA with each factor or IV having two levels. The IVs were sex (male and female) and meditate (used meditation and did not use meditation). The DV was the participants rating of their perceived level of stress measured on a scale of 1 (minimal stress) to 10 (high stress).

What were the degrees of freedom for this analysis? How did you reach this conclusion?

Total degrees of freedom:
$df_{tot} = N - 1 = 20 - 1 = 19$

Sex degrees of freedom:
$df_{sex} = k - 1 = 2 - 1 = 1$

Meditate degrees of freedom:

$df_{meditate} = k - 1 = 2 - 1 = 1$

Interaction (Sex*Meditate) degrees of freedom:

$df_{interaction} = (df_{sex})(df_{meditate}) = (1)(1) = 1$

Within or error degrees of freedom:

$df_{within} = N - (\text{# of participant sex groups})(\text{# of meditation groups}) = 20 - (2)(2) = 16$

Where:

N = the total number of perceived stress scores or participants for all groups
k = the number of IV levels or factors

Were the results of this analysis significant at the pre-set .05 alpha level? Should a post hoc test be conducted? Why or Why not?

For this example we obtained F ratios for both IVs (sex and meditate) as well as their interaction (see SF-15.8). The main effect of sex was significant at the .05 level ($p = .021$) and the main effect of meditate was significant at the .05 level ($p = .000$). The interaction effect of sex and meditate was also significant at the .05 level ($p = .001$). Remember when an interaction is significant, it overrides any main effects that we obtain, so we can disregard the main effects for this example. The reason the main effects are disregarded is that the main effect is contained within the interaction and if the main effect was interpreted in a straightforward manner, we would be guilty of over simplifying the effect of that IV. Essentially, remember that a significant interaction means that we must refer to both IVs to understand the results of our findings. Usually the simplest way to understand an interaction effect is to graph the results. So as you complete the remaining questions you should refer to screen figure 15.9.

A post hoc test was not necessary for this example for two reasons. First, the interaction effect was significant and superseded the main effects. The second reason pertains to our experimental design. Because this example design only has two levels for the IVs we did not need to use a post hoc test because we can simply compare the two means as in a two-group design (i.e., the larger mean is significantly higher than the lower mean).

How would you write the statistical results of this experiment?

The main effect of the use of meditation practices on the air traffic controllers' perceived level of stress was significant, $F(1, 16) = 31.565$, $p = .000$, $\eta^2 = .664$. The main effect of sex on the air traffic controllers' perceived level of stress was also significant, $F(1, 16) = 6.52$, $p = .021$, $\eta^2 = .290$. However, the main effects were qualified by a significant interaction effect between the use of meditation practices and the sex of the air traffic controllers, $F(1, 16) = 18.11$, $p = .001$, $\eta^2 = .531$. The graphical results of the interaction appear in screen figure 15.9.

How would you interpret these results in non-statistical language?

Overall, after reviewing screen figure 15.9, the point on the graph that appears to differ most represents the male air traffic controllers' perceived levels of stress in the non-meditation condition. This mean is considerably elevated when compared to the other conditions. Thus, we can conclude that the male air traffic controllers who did not use meditation practices perceived themselves as experiencing higher levels of stress ($M = 9.8$) than did female air traffic controllers who did not use meditation practices ($M = 5.8$), and male ($M = 4.0$) and female ($M = 5.0$) air traffic controllers who did use meditation practices.

Answers to the multiple choice, matching, and true/false items

Multiple Choice		Matching		True and False	
1.	A	1.	A	1.	T
2.	E	2.	K	2.	F
3.	E	3.	L	3.	T
4.	B	4.	F	4.	F
5.	D	5.	G	5.	T
6.	D	6.	D	6.	T
7.	B	7.	I	7.	F
8.	A	8.	E	8.	T
9.	D	9.	H	9.	F
10.	D	10.	J	10.	F
11.	D	11.	C	11.	T
12.	A	12.	B		
13.	A				
14.	B				
15.	D				
16.	C				
17.	D				
18.	C				
19.	C				
20.	A				
21.	B				
22.	C				

Chapter 16

Alternative Research Designs

Learning Objectives

By the end of this chapter you should be able to:

1. Identify and summarize the most adequate control procedures for promoting internal validity?

2. List and describe the three experimental designs recommended by Campbell and Stanley (1966) for controlling threats to internal validity.

3. Summarize the advantages and limitations of the Solomon four-group design.

4. Explain why the concept of internal validity is critical to experimental designs and statistical analyses.

5. Discuss why the use of single-case experimental designs were more popular in the past.

6. List and describe several situations in which single-case designs are required.

7. List and describe common single-case designs.

8. Explain why researchers should only change one variable at a time in single-case designs.

9. Discuss the arguments for and against using statistical analyses with single-case designs.

10. Summarize the benefits and limitations of the A-B, A-B-A, and the A-B-A-B designs.

11. Discuss how a single-case design may be used to disprove a theory.

12. List and describe several specific situations that require quasi-experimental designs.

13. List and describe the four threats to internal validity associated with the nonequivalent group design.

Expanded Outline

I.   Protecting Internal Validity Revisited
           Examining Your Experiment from the Inside

           Protecting Internal Validity with Research Designs
               Random Assignment

                   Random Selection
               Experimental Design
                   The Pretest-Posttest Control Group Design
                       Selection
                       History
                       Maturation
                       Testing
                       Statistical Regression
                       Interaction with Selection
                       Experimental Mortality
                       Instrumentation
                       Diffusion or Imitation of Treatments
                   The Solomon Four-Group Design
                   The Posttest-Only Control-Group Design

           Conclusion

II.   Review Summary

III.   Check Your Progress

IV.   Single-Case Experimental Designs
           History of Single-Case Experimental Designs

           Uses of Single-Case Experimental Designs
               Experimental Analysis of Behavior
           General Procedures of Single-Case Experimental Designs
               Repeated Measures
               Baseline Measurement
                   Baseline
               Changing One Variable at a Time

           Statistics and Single-Case Experimental Designs
               The Case Against Statistical Analysis
               The Case for Statistical Analysis

Representative Single-Case Experimental Designs
     A (baseline measurement)
     B (measurement during or after treatment)
     A-B Design

     A-B-A Design

     A-B-A-B Design
     Design and the Real World

     Additional Single-Case Designs

V.   Review Summary

VI.   Check Your Progress

VII.   Quasi-Experimental Designs

     History of Quasi-Experimental Designs
     Uses of Quasi-Experimental Designs
     Representative Quasi-Experimental Designs
          Nonequivalent Group Design

          Interrupted Time-Series Design

VIII. Review Summary

IX.   Check Your Progress

X.   Key Terms

XI.   Looking Ahead

Practice Exam

Multiple Choice

Identify the letter of the choice that best completes the statement or answers the question.

_____ 1. Threats to internal validity include

    a. confounding variables.   d. both a and b
    b. extraneous variables.   e. both b and c
    c. intrinsic variables.

_____ 2. Which type of validity examines whether the IV actually created any observable change in the DV?

    a. internal   c. intrinsic
    b. external   d. extrinsic

_____ 3. Which of the following is considered to be the most important property of an experiment?

    a. internal validity   c. feasibility
    b. generalizability   d. reliability

_____ 4. According to Campbell and Stanley (1966), "the most adequate all-purpose assurance of lack of initial biases between groups" is

    a. matched samples.   c. repeated measures.
    b. natural sets.   d. randomization.

_____ 5. Researchers can maximize the effectiveness of randomization as a control technique by using

    a. smaller participant groups.   c. shorter testing intervals.
    b. larger participant groups.   d. longer testing intervals.

_____ 6. _____ is related to the issue of internal validity; whereas the concept of _____ is more associated with external validity.

    a. Random assignment,   c. Random selection,
        random selection   natural assignment
    b. Random selection,   d. Random assignment,
        random assignment   natural selection

_____ 7.   Which experimental design was <u>not</u> recommended by Campbell and Stanley (1966) as being able to control threats to internal validity?

   a.   pretest-posttest control group   c.   Solomon two-group
   b.   Solomon four-group   d.   posttest only control group

_____ 8.   The pretest-posttest control design consists of

   a.   two randomly assigned participant groups.   d.   both a and b
   b.   two pre-tested participant groups.   e.   all of the above
   c.   one participant group receiving the IV.

_____ 9.   When researchers use a pretest and a posttest for both groups in a pretest-posttest control group design they are controlling for the effects of

   a.   history.   c.   testing.
   b.   maturation.   d.   all of the above

_____ 10. The main advantage of the Solomon four-group design over the pretest-posttest control group design relates to

   a.   external validity.   c.   reliability.
   b.   internal validity.   d.   testing bias.

_____ 11. Which problem is considered to be a limitation of the Solomon four-group design?

   a.   statistical regression   c.   instrumentation decay
   b.   history   d.   statistical analysis of the data

_____ 12. The necessary features for creating a posttest only control group design are

   a.   correlated assignment of participants.   d.   both a and c
   b.   random assignment of participants.   e.   both b and c
   c.   the use of a control group.

_____ 13. The most critical type of validity in experimental research is

   a.   external.   c.   intrinsic.
   b.   internal.   d.   content.

_____ 14. A physician is conducting a one year study of the effects of a new cancer drug on a patient in his clinic. Which experimental design will the physician likely use?

a. single case
b. Solomon four-group
c. pretest-posttest
d. posttest only

_____ 15. Researchers primarily attribute the decline of the single-case experimental design to the development of this statistical procedure:

a. correlation
b. *z*-test
c. standard deviations
d. ANOVA

_____ 16. According to research (Duke, 1965), single-case experimental designs are appropriate for situations where researchers

a. cannot find other participants.
b. can assume perfect generalizability.
c. are limited to observe behaviors.
d. b and c
e. all of the above

_____ 17. According to Hersen (1982), the characteristics of single-case designs include

a. repeated measures.
b. baseline measurement.
c. altering one variable at a time.
d. a and b
e. all of the above

_____ 18. In single-case experimental designs researchers should collect at least _____ observations during the baseline period in order to establish a trend in the data.

a. 3
b. 4
c. 5
d. 6

_____ 19. In a single-case design researchers should only change _____ variables(s) at a time when moving from one phase of the experiment to the next.

a. one
b. two
c. three
d. four

_____ 20. In a single-case design the standard notation of "A" refers to the _____; whereas the standard notation of "B" refers to the _____.

a. baseline, measurement during or after treatment
b. measurement during or after treatment, baseline
c. baseline, posttest
d. posttest, baseline

21. The most parsimonious single-case design allowing for a causal relation to be drawn is the

   a. A-B.   c. A-B-A.
   b. B-A.   d. A-B-A-B.

22. A nutritionist is studying whether caffeine influences her client's dietary behavior. Over the next few weeks she asks her client to adhere to the following regimen:

   week one: no caffeine allowed
   week two: caffeine allowed
   week three: no caffeine allowed
   week four: caffeine allowed

   What type of single-case design is she using?

   a. A only   c. A-B-A
   b. A-B   d. A-B-A-B

23. Which design is considered to be the <u>weakest</u> for inferring causality?

   a. A-B   c. A-B-A-B
   b. A-B-A   d. A-B-A-B-A

24. Which situation precludes using a design other than the A-B design?

   a. it is unethical to reverse a treatment   d. a and b
   b. it is impractical to reverse a treatment   e. all of the above
   c. it is impossible to reverse a treatment

25. When researchers are able to manipulate an IV and measure a DV, but <u>cannot</u> randomly assign their participants to groups, they must use a _____ design.

   a. true experimental   c. single-case
   b. quasi-experimental   d. correlational

26. Which situation requires a quasi-experimental design?

   a. evaluating previous life experiences   d. b and c
   b. evaluating retrospective studies   e. all of the above
   c. evaluating the effects of poverty

____ 27. The interrupted time-series design more closely resembles the _____ design.

    a. A-B
    b. A-B-A

    c. one-way ANOVA
    d. two-way ANOVA

____ 28. _____ is the <u>most</u> prominent threat to the internal validity of the interrupted time-series design.

    a. Maturation
    b. Instrumentation

    c. History
    d. Regression

____ 29. Peter is using a design that involves two participant groups that are <u>not</u> randomly assigned. He is also comparing a non-treatment comparison group with a treatment group. Which design is he likely using?

    a. interrupted time-series
    b. nonequivalent group

    c. A-B
    d. A-B-A

Matching

a. case-study

b. nonequivalent group design

c. internal validity

d. history

e. A

f. B

g. baseline

h. single-case experimental design

i. interrupted time-series design

j. A-B-A design

k. A-B design

_____ 1. a design using two or more groups that are <u>not</u> randomly assigned

_____ 2. single-case design in which you measure the baseline behavior, institute a treatment, and use a posttest

_____ 3. a measurement of behavior made under normal conditions, without an IV, often called a control condition

_____ 4. an experiment consisting of one participant

_____ 5. a symbol referring to the baseline measurement in a single-case design

_____ 6. a symbol referring to the treatment outcome in a single-case design

_____ 7. the primary threat to internal validity in the interrupted-time-series design

_____ 8. an observation method in which records are compiled about a single participant

_____ 9. a quasi-experimental design involving a single group of participants that includes repeated pre-treatment measures, an applied treatment, and repeated post-treatment measures

_____ 10. a single-case design consisting of a baseline measurement, a treatment, a post-test, and a return to the baseline condition

_____ 11. an evaluation examining whether your IV was the only possible explanation of the result(s) shown for your DV

True/False

Indicate whether the sentence or statement is true or false.

_____ 1. The concept of internal validity revolves around confounding and extraneous variables.

_____ 2. External validity is the most important property of an experiment.

_____ 3. Randomization is a major problem when using the Solomon four-group design.

_____ 4. Randomization is considered the most adequate control for ensuring internal validity.

_____ 5. Researchers should use at least 5 observations during baseline measurements.

_____ 6. The simplest case design that allows researchers to demonstrate a cause-and-effect relationship is the A-B-A design.

_____ 7. When researchers are not able to randomly assign their participants to groups, they may be forced to use quasi-experimental designs.

_____ 8. The interrupted time-series design is more closely related to the factorial design.

_____ 9. Statistical regression is considered the primary internal validity threat for the interrupted time-series design.

_____ 10. The time between testing intervals in an interrupted time-series design should be as short as possible.

_____ 11. It may not be necessary to use matched groups because random assignment can be used to equate the participant groups.

_____ 12. Random assignment is often used synonymously with the term random selection.

_____ 13. The notation for single-case designs is read from left to right to denote the passage of time.

Short Answer

1. According to Campbell and Stanley (1966), what is the most adequate control procedure used to promote internal validity? What are the advantages and disadvantages associated with this control procedure?

2. List and describe the three experimental designs recommended by Campbell and Stanley (1966) for controlling threats to internal validity.

3. Summarize the advantages and limitations of the Solomon four-group design.

4. Explain why the lack of a pretest condition in the posttest-only control group design does not make it less powerful than the pretest-posttest or Solomon four-group control designs.

5. Explain why the concept of internal validity is critical to experimental designs and statistical analyses.

6. Discuss why the use of single-case experimental designs were more popular in the past.

7. List and describe several situations in which single-case designs are required.

8. List and describe common single-case designs.

9. Explain why it is important for researchers to only change one variable at a time in single-case designs.

10. Discuss the arguments for and against using statistical analyses with single-case designs.

11. Summarize the benefits and limitations of the A-B, A-B-A, and the A-B-A-B designs.

12. Distinguish between quasi-experimental and ex post facto designs. Describe the problems that result when researchers cannot randomly assign participants to groups.

13. Discuss how a single-case design may be used to disprove a theory.

14. List and describe several situations that require quasi-experimental designs.

15. List and describe the four threats to internal validity associated with the nonequivalent group design.

Answers to the multiple choice, matching, and true/false items

Multiple Choice		Matching		True and False	
1.	D	1.	B	1.	T
2.	A	2.	K	2.	F
3.	A	3.	G	3.	F
4.	D	4.	H	4.	T
5.	B	5.	E	5.	F
6.	A	6.	F	6.	T
7.	C	7.	D	7.	T
8.	E	8.	A	8.	F
9.	D	9.	I	9.	F
10.	A	10.	J	10.	T
11.	D	11.	C	11.	T
12.	E			12.	F
13.	B			13.	T
14.	A				
15.	D				
16.	E				
17.	E				
18.	A				
19.	A				
20.	A				
21.	A				
22.	D				
23.	A				
24.	E				
25.	B				
26.	E				
27.	A				
28.	C				
29.	B				

Chapter 17

External Validity and Critiquing Experimental Research

Learning Objectives

By the end of this chapter you should be able to:

1. Define and distinguish between internal and external validity.

2. Explain why the existence of external validity is dependent on internal validity.

3. Discuss why the experimental processes used to ensure internal validity become disadvantages for obtaining external validity.

4. List and describe the four factors affecting external validity as presented by Campbell and Stanley (1966).

5. Identify which external validity threat typically occurs in repeated measures designs.

6. Explain why white rats and college students are considered threats to external validity.

7. Discuss why the theories developed by Sigmund Freud and Erik Erikson have been characterized as biased.

8. Summarize the significance of Robert Guthrie's book, *Even The Rat Was White*, to experimental research and psychology.

9. Recognize why Douglas Mook questioned the idea that all experiments should be designed to generalize to the real world in his article titled, *In Defense of External Invalidity*.

10. Define and explain the rationale for conducting programmatic research.

Expanded Outline

I.   External Validity: Generalizing Your Experiments to the Outside
           Internal Validity
           External Validity
           Generalization
           Population Generalization
           Environmental Generalization
           Temporal Generalization

           Threats to External Validity (Based on Methods)
           Interaction of Testing and Treatment
           Interaction of Selection and Treatment
           Reactive Arrangements
                Demand Characteristics
           Multiple-Treatment Interference

           Threats to External Validity (Based on Our Participants)
           The Infamous White Rat
                Comparative Psychology

           The Ubiquitous College Student
                Convenience Sampling

           The "Opposite" or "Weaker" or "Inferior" or "Second" Sex
           Even the Rats and Students Were White
           Even the Rats, Students, Women, and Minorities Were American
                Cross-Cultural Psychology
                Ethnocentric
      The Devil's Advocate: Is External Validity Always Necessary?

           Replication
           Replication with Extension

II.   Review Summary

III.   Check Your Progress

IV.   Critiquing Experimental Research
           Guidelines for Critiquing Research

V.   Review Summary

Practice Exam

Multiple Choice

Identify the letter of the choice that best completes the statement or answers the question.

_____ 1.   This type of validity is associated with whether your experiment is confounded or whether your IV is the only possible explanation for your finding.

     a. external   c. content
     b. internal   d. intrinsic

_____ 2.   _____ validity focuses on whether your experimental results apply to populations and situations that are different from your experiment.

     a. External   c. Content
     b. Internal   d. Intrinsic

_____ 3.   Which of the following is not considered to be a customary type of generalization?

     a. temporal   c. population
     b. environmental   d. historical

_____ 4.   Veronique is conducting a lab study in which the student participant groups are learning a list of words based on three different instructional strategies. She wonders whether this lab study will have any relevance to learning in real world classrooms. What type of generalization is she contemplating?

     a. temporal   c. educational
     b. environmental   d. historical

___ 5. Which of the following is considered to be the most obvious threat to external validity in the pretest-posttest control group design?

    a. interaction of selection and treatment
    b. reactive arrangements

    c. multiple-treatment interference
    d. the interaction of testing and treatment

___ 6. A _____ occurs when the effects of your research results only apply to a particular participant group that you selected for your study.

    a. reactive arrangement
    b. selection-treatment interaction

    c. testing-treatment interaction
    d. multiple-treatment interaction

___ 7. The ability to generalize research findings beyond a certain time period best describes _____ generalization.

    a. longitudinal
    b. durational

    c. true
    d. temporal

___ 8. The conditions of an experimental setting, other than the IV(s), that alter the behavior of the participants in the study is referred to as

    a. reactive arrangements.
    b. multiple-treatment interactions.

    c. selection and treatment interactions.
    d. testing and treatment interactions.

___ 9. This type of generalization is concerned with applying research results beyond the researcher's experimental participants.

    a. historical
    b. population

    c. environmental
    d. extended

___ 10. Linda conducted an experiment and learned that the effects of her study only applied to college students. What type of internal validity threat has she identified?

    a. reactive arrangements
    b. multiple-treatment interactions

    c. interaction of selection and treatment
    d. interaction of testing and treatment

___ 11. _____ provide clues about the experimental hypothesis and how to respond to stimuli.

    a. Reactive arrangements
    b. Biased questionnaires

    c. Correlated groups
    d. Demand characteristics

_____ 12. The multiple-treatment interference threat to external validity typically occurs in _____ designs.

   a. repeated measure   c. reactive arrangement
   b. testing-treatment interaction d. selection-treatment interaction

_____ 13. Which animal dominated experimental research in the 1930s?

   a. pigeons   c. mice
   b. white rats   d. monkeys

_____ 14. In Robert Guthrie's book entitled _____, he chronicled many of the "scientific" attempts to measure and categorize African-Americans as inferior to Caucasians.

   a. _The Infamous White Rat_   c. _Even The Rat Was White_
   b. _The Eight Stages of Man_   d. _The Ubiquitous African-American_

_____ 15. Steve has just completed his senior research project. He is curious if his experimental results would apply to people in other countries. What concept should he consider?

   a. external validity   d. a and c
   b. cross-cultural psychology e. all of the above
   c. generalizability

_____ 16. Researchers who retest for a particular experimental finding in a different context are using

   a. extended designs.   c. replication with extension.
   b. replication.   d. longitudinal studies.

_____ 17. Which researchers developed theories that were considered to be biased toward men?

   a. Tavris and Freud   c. Tavris and Guthrie
   b. Erikson and Guthrie d. Freud and Erikson

_____ 18. While critiquing experimental research, why should researchers be concerned if the hypothesis is stated in general implication form?

   a. it improves readability   d. adheres to APA style
   b. it allows for cause-and-effect e. all of the above
      conclusions to be drawn.
   c. adheres to ethical guidelines

____ 19. When reviewing the introduction section of a research report, you should closely examine the

    a. literature review.   d. citations and references.
    b. research question.   e. all of the above
    c. hypothesis.

____ 20. Sally asks her friend to critique the results section of her paper. What content should Sally's friend expect to find?

    a. descriptive statistics   d. both a and b
    b. inferential statistics   e. both b and c
    c. the hypothesis statement

____ 21. What information would you expect to find in the methods section of a research paper?

    a. inferential statistics   c. references
    b. description of the participants   d. conclusions

Matching

a. external validity
b. Francis Cecil Sumner
c. Carol Tavris
d. programmatic research
e. Frank Beach

f. Robert Guthrie
g. internal validity
h. Sigmund Freud
i. Douglas Mook
j. demand characteristics

_____ 1. research that extends or builds upon previous experiments

_____ 2. conveys the experimental hypothesis to the participants and provides clues about how to respond to stimuli

_____ 3. examines whether your IV is the only possible explanation of the results shown for your DV

_____ 4. examines whether your results generalize beyond your original experiment

_____ 5. author of *Even The Rat Was White*

_____ 6. identified that white rats dominated research in the 1930s

_____ 7. has been labeled as the "Father of Black American Psychology"

_____ 8. questioned whether we are developing a body of knowledge pertaining to all organisms, regardless of sex

_____ 9. this researcher's theories have been labeled as biased toward men

_____ 10. author of *In Defense of External Invalidity*

True/False

Indicate whether the sentence or statement is true or false.

_____ 1. Researchers <u>only</u> worry about evaluating their experiment for external validity after they have completed their experiment.

_____ 2. External validity can exist independently from internal validity.

_____ 3. According to Orne (1962), it is impossible to design an experiment without demand characteristics.

_____ 4. Pigeons were the most frequently used research animals in the 1930s.

_____ 5. Convenience sampling is most associated with college students.

_____ 6. A threat to internal validity that occurs when a pretest sensitizes participants to the upcoming treatment is known as the interaction of testing and treatment.

_____ 7. An experiment that seeks to confirm a previous finding, but does so in a different context is called replication.

_____ 8. Programmatic research involves a series of experiments focusing on a related topic.

_____ 9. Generalization is more relevant to internal validity than external validity.

_____ 10. Internal validity examinations are concerned with whether experimental results apply to populations and situations that are different from those of the current experiment.

Short Answer

1. Define and distinguish between the terms internal and external validity. Explain the relationship between these types of validity.

2. Explain why the existence of external validity is dependent on internal validity. Provide a brief example to demonstrate this relationship.

3. Summarize why the experimental processes used to ensure internal validity become disadvantages for obtaining external validity.

4. List and describe the four factors affecting external validity as presented by Campbell and Stanley (1966). Which of the four is the most obvious threat to external validity and occurs for the pretest-posttest control group design?

5. Identify the threats to external validity that typically occur in repeated measures designs. Explain the potential problem that may arise with this design.

6. Summarize why animal research (Beach, 1950; Smith, Davis, & Burleson, 1995) has provided concern regarding external validity.

7. Explain why white rats and college students can be threats to external validity. Are these participant groups considered to be convenience samples?

8. Discuss why the theories developed by Sigmund Freud and Erik Erikson have been characterized as biased.

9. Summarize the significance of Robert Guthrie's book, *Even The Rat Was White*, to experimental research and psychology.

10. Explain why Douglas Mook attacked the idea that all experiments should be designed to generalize to the real world in his article titled, *In Defense of External Invalidity*.

11. Define programmatic research. Explain the rationale for conducting this type of research.

Answers for the multiple choice, matching, and true/false items

Multiple Choice		Matching		True and False	
1.	B	1.	D	1.	F
2.	A	2.	J	2.	F
3.	D	3.	G	3.	T
4.	B	4.	A	4.	F
5.	D	5.	F	5.	T
6.	B	6.	E	6.	F
7.	D	7.	B	7.	F
8.	A	8.	C	8.	T
9.	B	9.	H	9.	F
10.	C	10.	I	10.	F
11.	D				
12.	A				
13.	B				
14.	C				
15.	E				
16.	C				
17.	D				
18.	B				
19.	E				
20.	D				
21.	B				

Chapter 18

Data Transformations and Nonparametric Tests of Significance

Learning Objectives

By the end of this chapter you should be able to:

1. List and describe the assumptions for conducting parametric tests.

2. Explain the options researchers have after violating an assumption for conducting parametric tests.

3. Discuss what test is used to examine the assumption that the population variances are equivalent.

4. Summarize how data transformations change the characteristics of a distribution.

5. List several methods that are used to transform data.

6. Compare and contrast chi-square tests for two or more levels of a single nominal variable.

7. Discuss the purpose for using rank-order tests.

8. Explain what situation(s) are appropriate for using the Mann-Whitney U, the rank sums, the Kruskal-Wallis, and the Wilcoxon tests.

By the end of this chapter you should learn when and how to use the following formulas:

I. The formula for a chi-square test for two levels of a single nominal variable

$$\chi_{obt}^{2} = \Sigma \left[\frac{(O-E)^{2}}{E} \right]$$

(1)

Where:

O = the frequency of observed scores
E = the frequency of the expected scores

- At this point it may be helpful to view a matrix for this data comparison

	Category or Group 1	Category or Group 2
Expected Frequencies	f_{e_1}	f_{e_2}
Observed Frequencies	f_{o_1}	f_{o_2}

Where:

f_{e_1} =   expected frequencies for group 1

f_{e_2} =   expected frequencies for group 2

f_{o_1} =   observed frequencies for group 1

f_{o_2} =   observed frequencies for group 2

- The steps for calculating a chi-square test for two levels of a single nominal variable

1. Subtract the appropriate expected value (f_e) from each corresponding observed value (f_o). Then square the differences and divide the squared values by the expected values.

2. Sum all of the products calculated in step 1. Steps 1 and 2 can be completed by using formula 2.

$$\chi_{obt}^2 = \left[\frac{\left(f_{o_1} - f_{e_1}\right)^2}{f_{e_1}} \right] + \left[\frac{\left(f_{o_2} - f_{e_2}\right)^2}{f_{e_2}} \right]$$

(2)

3. Next, calculate the degrees of freedom.

$df = k - 1$

Where:

k =   the number of categories or groups

4. Lastly, look up the <u>chi-square (χ^2) critical value</u> for your *df* at the appropriate significance level (i.e., .05 or .01). If your chi-square obtained value (χ^2_{obt}) is larger than the chi-square critical value for the corresponding significance level, you can conclude that your result was significant.

II. <u>The formula for a chi-square test with two nominal variables</u>

$$\chi^2_{obt} = \left[\frac{\left(f_{o_1}-f_{e_1}\right)^2}{f_{e_1}}\right] + \left[\frac{\left(f_{o_2}-f_{e_2}\right)^2}{f_{e_2}}\right]$$

(2)

- <u>The steps for calculating the test statistic for a chi-square test with two nominal variables</u>

1. Obtaining the expected frequency for any of the cells in your contingency table for two nominal variables requires using formula 4. Once the expected frequency value (f_e) is obtained you can plug this value into formula 2, which is the same formula used for two levels of a single nominal variable.

$$f_e = \frac{(RT)(CT)}{N}$$

(3)

Where:

RT = the row total
CT = the column total
N = the grand total

- For a 2 x 2 contingency table, once you have obtained the expected frequency for one cell you can then calculate the remaining cells by utilizing the row and column totals and subtracting.

2. Next, calculate the <u>degrees of freedom.</u>

df = (*number of rows* – 1) (*number of columns* – 1)

3. Lastly, look up the <u>chi-square (χ^2) critical value</u> for your *df* at the appropriate significance level (i.e., .05 or .01). If your chi-square obtained value (χ^2_{obt}) is larger than the chi-square critical value for the corresponding significance level, you can conclude that your result was significant.

III. The formula for calculating the Mann-Whitney U test

$$U_1 = (n_1)(n_2) + \left[\frac{n_1\,(n_1 + 1)}{2}\right] - \Sigma R_1$$

(4)

and

$$U_2 = (n_1)(n_2) + \left[\frac{n_2\,(n_2 + 1)}{2}\right] - \Sigma R_2$$

(5)

Where:

ΣR_1	=	the sum of the ranks for group 1
ΣR_2	=	the sum of the ranks for group 2
n_1	=	the number of categories or participants in group 1
n_2	=	the number of categories or participants in group 2

- The steps for calculating the Mann-Whitney U test

1. Rank all of the scores in your data set without considering group membership. The lowest score will be assigned the rank of 1.

2. Sum the ranks for each group (i.e., ΣR_1 and ΣR_2).

3. Compute a U value for each group (i.e., U_1 and U_2) using formulas 4 and 5.

4. Determine which U value (i.e., U_1 or U_2) you will use to test for significance. If you are using a two-tailed test (non-directional), you will use the smaller U value; whereas if you are using a one-tailed test (directional), you will use the U value for the group you are predicting will have the larger sum of ranks.

5. Next, obtain the critical U value from the appropriate table.

6. Lastly, compare your calculated U value to the critical value. To be significant, your calculated U value must be equal to or less than the table value.

IV. The formula for calculating the rank sums test

$$z_{obt} = \frac{\Sigma R - \Sigma R_{exp}}{\sqrt{\dfrac{(n_1)(n_2)(N+1)}{12}}}$$

(6)

Where ΣR_{exp} can be obtained using formula 7:

$$\Sigma R_{exp} = \frac{n(N+1)}{2}$$

(7)

• The steps for calculating the rank sums test

1. Rank order all of the scores in your data set without considering group membership. The lowest score will be assigned the rank of 1.

2. Select one group and sum the ranks for that group (ΣR).

3. Using formula 7 calculate the expected sum of ranks (ΣR_{exp}).

4. Then use the ΣR and ΣR_{exp} values for the selected group to calculate the rank sums z_{obt} statistic using equation 6.

5. Next, use the appropriate table to determine the critical z value for the selected alpha level (i.e., .01 or .05). Remember if you are using a two-tailed test (non-directional) you will equally split the alpha level; hence for the .05 alpha level you will test at the .025 level in both tails. If your obtained rank sums absolute value (z_{obt}) is larger than the critical value, then your groups significantly differ. Remember to use the absolute value of your calculated z_{obt} statistic when comparing it with the critical table value.

6. Lastly, calculate an effect size (ex., eta squared) by using formula 8:

$$\eta^2 = \frac{(z_{rank\,scores})^2}{12-1}$$

(8)

V. The formula for calculating the Wilcoxon T test

$$T_{obt} = \Sigma R$$

254

- The steps for calculating the Wilcoxon *T* test

1. Obtain the difference between the scores for each pair. It does not matter which score is subtracted as long as you are consistent throughout all the pairs.

2. Assign ranks to all nonzero differences and disregard the sign of the difference. The smallest difference will be assigned a rank of 1. Tied differences receive the average of the ranks they are tied with.

3. Separately determine the ranks that are based on positive and negative differences.

4. Sum the positive and negative ranks. These sums are the *T* values you will use to determine significance.

5. If you are using a non-directional hypothesis (i.e., two-tailed), you will use the smaller of the two sums of ranks; whereas if you have a directional hypothesis (i.e., one-tailed) you will have to determine which sum of ranks your hypothesis predicts will be smaller.

6. The number of paired groups (i.e., *N*) for the Wilcoxon *T* test will equal the number of <u>nonzero</u> differences.

7. Lastly, compare your critical value with the obtained *T* value. For this value to be significant the obtained value (T_{obt}) must be equal to or less than the table value.

VI.   <u>The formula for calculating the Kruskal-Wallis *H* test</u>

$$H_{obt} = \left[\frac{12}{N\,(N+1)} \right] (SS_{BG}) - 3(N+1)$$

(9)

Where SS_{BG} can be obtained using formula 10:

$$SS_{BG} = \frac{(\Sigma R_1)^2}{n_1} + \frac{(\Sigma R_2)^2}{n_2} + \cdots \frac{(\Sigma R_k)^2}{n_k}$$

(10)

Where:

SS_{BG}	=	sum of squares between groups
n	=	group or condition
N	=	total number of groups or conditions
ΣR	=	the sum of the ranks for a group or condition

- <u>The steps for calculating the Wilcoxon T test</u>

1. Rank all of the scores with the lowest score equaling 1.

2. Sum the ranks for each group or condition.

3. Square the sum of the ranks for each group or condition.

4. Use formula 10 to calculate the sum of squares between groups (SS_{BG}).

5. Once you have obtained the SS_{BG} you can then plug this value into formula 9 to calculate the H obtained (H_{obt}) value.

6. Use the appropriate chi-square table to find the critical value for H. If the obtained H value is greater than the table value the result is significant.

7. Calculate the <u>degrees of freedom</u> using formula.

   $df = k - 1$

8. Calculate an <u>effect size</u> (ex., eta squared) by using formula 11.

$$\eta^2 = \frac{H}{N-1} \tag{11}$$

9. Lastly, if appropriate conduct a <u>post hoc test</u>.

Expanded Outline

I. Assumptions of Inferential Statistical Tests and the F_{max} Test
Parametric Tests
Homogeneity of Variance
F_{max} Test

II. Data Transformations
Common Data Transformations
Square Root
Logarithmic
Reciprocal
What Data Transformations Do and Do Not Do
Using Data Transformations and a Caveat

III. Nonparametric Tests of Significance

IV. Chi-Square Tests
Chi-Square Test for Two Levels of a Single Nominal Variable
Expected Frequency Distribution
Observed Frequency Distribution
Calculating the Chi-Square Statistic
Chi-Square Test for More Than Two Levels of a Single Nominal Variable

Chi-Square Test for Two Nominal Variables

Contingency Table
Chi-Square Tests for Independence

V. Review Summary

VI. Check Your Progress

VII. Rank-Order Tests
Rationale for Rank-Order Tests
Mann-Whitney U Test
Rank Sums Test

Wilcoxon T Test
Kruskal-Wallis H Test

VIII. Review Summary

Practice Exam

Multiple Choice

Identify the letter of the choice that best completes the statement or answers the question.

_____ 1.   An inferential test of significance that estimates population parameters is called a(n)

    a. non-parametric test.   c. parametric test.
    b. F_{max} Test.   d. contingency table.

_____ 2.   Which of the following are assumptions that researchers make when conducting parametric tests?

    a. the population samples are skewed   d. both a and b
    b. the population samples are normally distributed   e. both b and c
    c. the population samples are homogeneous

_____ 3.   A(n) _____ is used to test the assumption that population variances are equivalent.

    a. *t*-test   c. *p*-value
    b. F_{max} d. chi-square test

_____ 4.   All of the following are common data transformations <u>except</u> for the:

    a. square root   c. logarithm
    b. chi-square   d. reciprocal

_____ 5.   The chi-square test is considered a

    a. parametric test.   c. nonparametric test.
    b. distribution-dependent test.   d. descriptive statistic.

_____ 6.   The chi-square test ordinarily uses _____ data.

   a. nominal   c. interval
   b. ordinal   d. ratio

_____ 7.   A chi-square test of two categorical variables is called a

   a. parametric test.   c. contingency test.
   b. goodness-of-fit test.   d. test of independence.

_____ 8.   If the χ^2 obtained value is less that the χ^2 critical or table value, the differences between expected and observed frequencies can be attributed to

   a. the test effect   c. bias
   b. chance fluctuation   d. true differences

_____ 9.   In a 2 x 2 contingency table, the sum of the expected frequencies for any row or column is

   a. less than the respective   c. equal to the respective
       row or column total.   row or column total.
   b. more than the respective   d. none of the above
       row or column total.

_____ 10.   Researchers calculate the df for a chi-square test of independence by using

   a. (row - 1) (column - 1).   c. (row - 1).
   b. (column - 1).   d. $N - 1$.

_____ 11.   When using a chi-square test of independence, the expected frequency for a cell is the product of the marginal totals divided by

   a. the row total.   c. N.
   b. the cell total.   d. $N - 1$.

_____ 12.   The test statistic for the Mann-Whitney test is

   a. z c. E
   b. U d. χ^2

____ 13. When researchers use the Mann-Whitney test for small samples and find that the obtained value is greater than the critical value of the test statistic they should conclude that

    a. the null hypothesis is not rejected.   c. a large sample version should be used.
    b. the null hypothesis is rejected.   d. the median test should be used.

____ 14. Researchers utilize the Mann-Whitney U as a nonparametric alternative for the

    a. chi-square.   c. the two-sample t-test for dependent samples.
    b. the F-ratio.   d. the two-sample t-test for independent samples.

____ 15. A nonparametric test that requires that the obtained value be smaller than the critical table value in order to reject the null hypothesis is the

    a. Tukey test.   c. rank sums test.
    b. 2 x 2 chi-square.   d. Mann-Whitney U test.

____ 16. If a researcher has used _____ independent group(s) <u>and</u> the n in one or both groups is _____ than 20, then the rank sums test is the appropriate ordinal data statistic.

    a. one, smaller   c. two, smaller
    b. one, greater   d. two, greater

____ 17. The first step for calculating the Wilcoxon test involves

    a. ranking the combined scores.   c. summing the scores in each group.
    b. finding the difference between   d. ranking the difference scores.
       the scores for each pair.

____ 18. The null hypothesis associated with the Kruskall-Wallis test is that

    a. there is no difference in the   c. there is no difference in the
       scores of the k populations.   means of the k populations.
    b. there is no difference in the   d. none of the above
       medians of the k populations.

____ 19. The test statistic for the Kruskal-Wallis one-way analysis of variance test is:

    a. T c. χ^2
    b. U d. H

_____ 20.   The test statistic for the Wilcoxon test is:

a. T c. χ^2

b. U d. H

Matching

a. reciprocal   g. nonparametric tests

b. χ^2 h. rank order

c. observed frequency distribution   i. parametric tests

d. H j. F_{max}

e. contingency table   k. U

f. expected frequency distribution

_____ 1.   the anticipated distribution of frequencies into categories

_____ 2.   the test statistic for the Kruskal-Wallis one-way analysis of variance

_____ 3.   a test for the equality of variance in which the largest variance is divided by the smallest variance

_____ 4.   a data transformation method

_____ 5.   a nonparametric test appropriate for ordinal data

_____ 6.   inferential tests of significance that attempt to estimate population parameters

_____ 7.   organizes the distribution of frequencies for two nominal variables

_____ 8.   the test statistic for the chi-square test

_____ 9.   significance tests that do not attempt to estimate population parameters such as means and variances

_____ 10.   the test statistic for the Mann-Whitney test

_____ 11.   the actual distribution of frequencies into categories

True/False

Indicate whether the sentence or statement is true or false.

_____ 1.   A nonparametric test of significance estimates population parameters.

_____ 2.   A chi-square tests the assumption that the population variances are equivalent.

_____ 3.   Data transformations are used to normalize distributions.

_____ 4.   Data transformations alter the relative position of the data in the sample.

_____ 5.   Researchers use nonparametric tests with nominal and ordinal data.

_____ 6.   A parametric test is an inferential test that estimates population parameters.

_____ 7.   The Kruskal-Wallis H test is analogous to a t-test for correlated groups.

_____ 8.   The Wilcoxon T test is used to analyze ordinal data for two unrelated groups.

_____ 9.   A rank-order test is a nonparametric test that is appropriate for ordinal data.

_____ 10.   The Mann-Whitney test is used for two relatively large independent groups.

Short Answer

1.   List and describe the assumptions for conducting parametric tests.

2.   Explain the options researchers have after violating an assumption for conducting parametric tests.

3.   Discuss what test is used to examine the assumption that the population variances are equivalent. List and describe the steps for conducting this test.

4.   Summarize how data transformations change the characteristics of a distribution. List several methods that are used to transform data.

5.   Compare and contrast chi-square tests for two or more levels of a single nominal variable. Are there differences in how the expected frequencies are obtained? Explain your answer.

6.   Discuss the purpose for using rank-order tests.

7.   Explain what situation(s) are appropriate for using the Mann-Whitney U test, the rank sums test, the Kruskal-Wallis test, and the Wilcoxon test.

SPSS Computer Practice Problem 18.1 Using SPSS 11.0 for Windows

Basically when conducting a statistical analysis using SPSS for Windows you will name your variables, enter your data, and finally analyze your data by selecting options from a toolbar. In the following section a step by step procedure of an example problem will be provided to guide you in calculating a chi-square test for two nominal variables. By following the systematic instructions provided and by referring to the screen figures (SF) when prompted, you should be able to independently conduct a chi-square test for two nominal variables or a chi-square test for independence.

Data

The fictional data for this problem is based on a psychology teacher who examined whether a research methods course was a necessary prerequisite for her statistics course. She recorded the number of students passing and failing the course as a function of whether they had successfully completed a research methods course over the past few semesters. The contingency table and the raw data for this example are provided below.

Contingency Table

Student Group	Statistics Course Outcome	
	Pass (1)	Fail (2)
Research methods taken (1)	14	6
Research methods not taken (2)	7	13

Raw Data Table

Number of Students	Student Group	Statistics Course Outcome
1	1.00	1.00
2	1.00	1.00
3	1.00	2.00
4	2.00	1.00
5	2.00	1.00
6	2.00	2.00
7	2.00	2.00
8	1.00	1.00
9	1.00	1.00
10	1.00	1.00
11	1.00	2.00
12	2.00	1.00
13	2.00	2.00
14	2.00	2.00
15	1.00	1.00
16	1.00	2.00
17	2.00	1.00
18	2.00	2.00
19	2.00	2.00
20	1.00	1.00
21	1.00	2.00
22	2.00	1.00
23	2.00	2.00
24	1.00	1.00
25	1.00	2.00
26	2.00	2.00
27	1.00	1.00
28	1.00	1.00
29	2.00	2.00
30	1.00	1.00
31	2.00	1.00
32	1.00	1.00
33	2.00	2.00
34	1.00	1.00
35	2.00	2.00
36	2.00	2.00
37	1.00	1.00
38	2.00	1.00
39	1.00	2.00
40	2.00	2.00

While working on this example problem consider the following questions:

1. What are the degrees of freedom for this analysis?

2. Are the results of this analysis significant at the pre-set .05 alpha level?

3. How will you write the statistical results of this experiment?

4. How will you interpret the results in non-statistical language?

Procedures for Completing Example Problem 18.1

1. **Data Entry.** This example will have data for two nominal variables. The first variable, "group," will include information about students who did ("1") or did not take a research methods course ("2") prior to taking their statistics course. In sum, 20 of the 40 students took a research methods course prior to taking the statistics course (see the contingency table). The second variable, "outcome," will include information about students who passed ("1") or failed the statistics course ("2"). In sum, 21 of the 40 students passed the statistics course (see the contingency table). For example, the data for participant number 20 in screen figure 18.1, will have taken the research methods course (group = "1") and their corresponding score explains that they passed the statistics course (outcome = "1"). To conduct this analysis you will need to enter the raw data for all of the student participants ($N = 40$) as provided in the raw data table.

	group	outcome	var	var	var	var	var	var	var	var
20	1.00	1.00								
21	1.00	2.00								
22	2.00	1.00								
23	2.00	2.00								
24	1.00	1.00								
25	1.00	2.00								
26	2.00	2.00								
27	1.00	1.00								
28	1.00	1.00								
29	2.00	2.00								
30	1.00	1.00								
31	2.00	1.00								
32	1.00	1.00								
33	2.00	2.00								
34	1.00	1.00								
35	2.00	2.00								
36	2.00	2.00								
37	1.00	1.00								
38	2.00	1.00								
39	1.00	2.00								
40	2.00	2.00								

SF-18.1. Entering data to conduct a chi-square test for two nominal variables.

Before conducting the chi-square test for two nominal variables it may be helpful to label the values you have used in your "group" and "outcome" variables. This will make your output much easier to interpret. Labeling these values can be accomplished by selecting the "Variable View" button on the bottom left corner of the data entry screen (see SF-18.1). Once selected your screen should resemble screen figure 18.2. Next click on the box below the "Values" column for row one. This box should have the word "None" in it before selected. Once selected a "Value Labels" command box will appear (see SF-18.2). At this point enter a "1" in the "Value" cell and enter the title ("research methods taken") of this value in the "Value Label" cell. Then select the "Add" button in the lower left hand corner of the "Value Labels" command box to move this information into the larger white box. If you make a mistake you can use the "Remove" command and start over. Repeat this process for the second value of the "group" variable ("research methods not taken"). Then label the values for the "outcome" variable (i.e., "pass" = 1; "fail" = 2). Once you have finished select the "OK" button on the "Value Labels" command box and return to the data entry screen by selecting "Data View" at the bottom left hand corner of the screen (see SF-18.2).

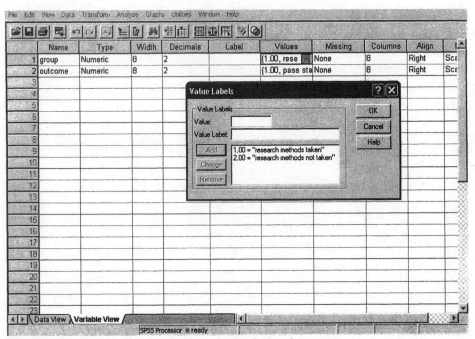

SF-18.2. Entering data: Labeling the coded values.

2. **Conducting the chi-square test for two nominal variables.** In order to conduct a statistical analysis using SPSS you will need to select the "Analyze" toolbar option at the top of the screen. A dropdown box will appear providing further data analysis options. Select the "Descriptive Statistics" option. After selecting this option SPSS will provide you with further analysis alternatives to the right of the initial dropdown box (see SF-18.3). At this point select the "Crosstabs" option.

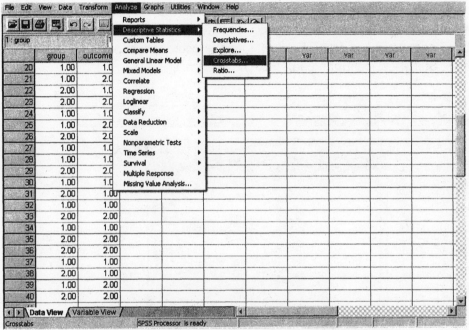

SF-18.3. Selecting the chi-square test for two nominal variables.

Once you have selected the "Crosstabs" option you should be able to view a command box named "Crosstabs" (see SF-18.4). At this point the variables "group" and "outcome" will appear in the large white box on the left hand side of the "Crosstabs" command box. Move the "group" variable into the "Row" box by first highlighting (it should appear blue) this variable and then by selecting the arrow next to the "Row(s):" box (see SF-18.4). Next, move the "outcome" variable into the "Column(s):" box by selecting the arrow button closest to the "Column(s):" box. At this point you will need to make a decision regarding which crosstab procedure you want to use for your analysis. This can be accomplished by selecting the "Statistics" button on the bottom of the "Crosstabs" command box (see SF-18.4). Once selected you should be able to view a "Crosstabs: Statistics" command box (see SF 18.5).

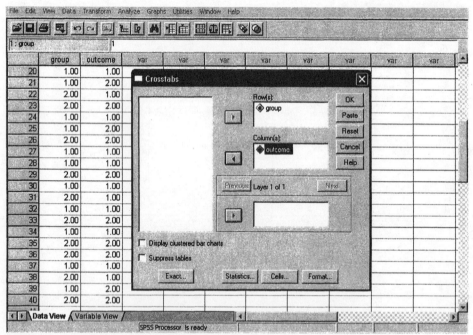

SF-18.4. Conducting the chi-square test for two nominal variables.

At this point you should be able to view the "Crosstabs: Statistics" command box. Select the "Chi-square" option in the upper left hand corner of the "Crosstabs: Statistics" command box and then select the "Continue" command. This will return you to the "Crosstabs" command box. Now you will need to make a few more decisions before conducting your analysis. Your screen should once again resemble screen figure 18.4. At this point select the "Cells" button located on the bottom middle portion the "Crosstabs" command box (see SF-18.4). Once selected your screen should resemble screen figure 18.6.

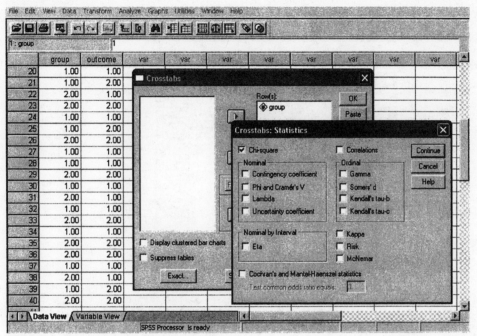

SF-18.5. Selecting a chi-square test.

For the purpose of this example you will need to select the "Observed" and "Expected" options under the "Counts" heading in the "Crosstabs: Cell Display" command box (see SF-18.6). These commands will prompt SPSS to include both the observed and expected group values in your output. Now select the "Continue" button in the upper right hand corner of the "Crosstabs: Cell Display" command box and SPSS will return you to the "Crosstabs" command box. Finally, select the "OK" button in the upper right hand corner of the "Crosstabs" command box and this will prompt SPSS to conduct your analysis.

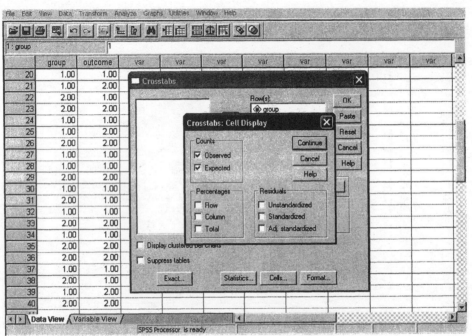

SF-18.6. Selecting options for your chi-square test.

3. **Reviewing the Output for the chi-square test for two nominal variables.** Your output should resemble the output in SF-18.7. If it does not you may want to check your raw data for data entry errors. The first output labeled "Case Processing Summary" provides information about the compared variables (i.e., "GROUP*OUTCOME") and the number of valid and missing cases. The second output box labeled "GROUP*OUTCOME" Crosstabulation" provides information regarding the observed ("Count") and expected frequency counts ("Expected Count") for the variables.

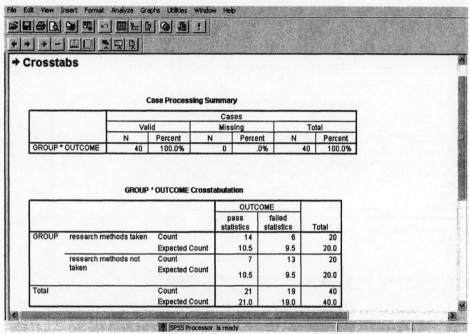

SF-18.7. Output for the chi-square test: The crosstabulation table.

If you scroll the output down you will find the information for your chi-square test (see SF-18.8). The "Chi-Square Tests" table provides a great deal of information. However, for this example, you only need to attend to the "Pearson Chi-Square" output, which is the ordinary chi-square test. This table provides the obtained chi-square statistic ($\chi^2 = 4.912$), the degrees of freedom ($df = 1$), and the significance value ($p = .027$).

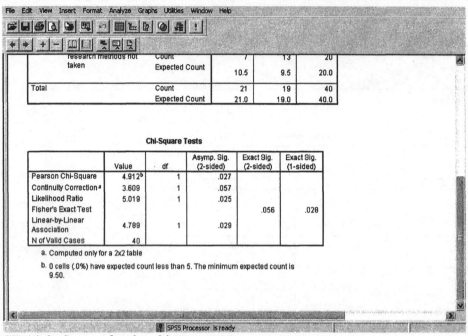

Chi-Square Tests

	Value	df	Asymp. Sig. (2-sided)	Exact Sig. (2-sided)	Exact Sig. (1-sided)
Pearson Chi-Square	4.912[b]	1	.027		
Continuity Correction[a]	3.609	1	.057		
Likelihood Ratio	5.019	1	.025		
Fisher's Exact Test				.056	.028
Linear-by-Linear Association	4.789	1	.029		
N of Valid Cases	40				

a. Computed only for a 2x2 table

b. 0 cells (.0%) have expected count less than 5. The minimum expected count is 9.50.

SF-18.8. Output for the chi-square test.

4.   **Interpreting the Results**.

What were the degrees of freedom for this analysis? How did you reach this conclusion?

The degrees of freedom for this analysis can be calculated by:

df = (number of rows – 1) (number of columns – 1)

df = (2-1) (2-1) = (1) (1) = 1

Were the results of this analysis significant at the pre-set .05 alpha level?

Yes the result was significant [χ^2 (1, N = 40) = 4.912, p = .027] given that the p-value was less than the pre-set significance level of .05 (see SF-18.8).

How would you write the statistical results of this experiment?

For this example we have two distributions: the distribution of students who passed or failed their statistics course (Outcome) and the distribution of the students who did or did not take a research methods course prior to taking their statistics course (Group). Remember the chi-square for a contingency table tests to determine if the distributions in the table (Outcome by Group) have similar or different patterns. Because our chi-square was significant, the group distributions are considered to be dissimilar or dependent. For this example, because the distributions are considered to be dependent, the pattern for the students taking a research methods course prior to taking their statistics course is <u>not</u> the same as the pattern for the students who did not take a research methods course prior to taking their statistics course. In other words, the nature of the pattern of the nominal "outcome" variable depended on the specific category of the nominal "group" variable. The results can be written as:

> The distribution of patterns for those students passing or failing their statistics course depended on whether the students previously took a research methods course χ^2 (1, N = 40) = 4.912, p = .027.

How would you interpret these results in non-statistical language?

Practically speaking, the psychology teacher can conclude that passing her statistics course depended on whether her students had or had not taken a research methods course.

Answers for the multiple choice, matching, and true/false items

Multiple Choice		Matching		True and False	
1.	C	1.	F	1.	F
2.	E	2.	D	2.	F
3.	B	3.	J	3.	T
4.	B	4.	A	4.	F
5.	C	5.	H	5.	T
6.	A	6.	I	6.	T
7.	D	7.	E	7.	F
8.	B	8.	B	8.	F
9.	C	9.	G	9.	T
10.	A	10.	K	10.	F
11.	C	11.	C		
12.	B				
13.	A				
14.	D				
15.	D				
16.	D				
17.	B				
18.	A				
19.	D				
20.	A				

Chapter 19

Nonexperimental Methods

Learning Objectives

By the end of this chapter you should be able to:

1. Discuss the purpose for using descriptive research methods.

2. List and describe the most common observational techniques used by researchers.

3. Explain why researchers should be unobtrusive when using naturalistic observation research methods.

4. Contrast and compare the limitations and benefits associated with using participant observations.

5. Distinguish between the clinical perspective and participant observation research methods.

6. Summarize the purpose for using interrater reliability measures.

7. Explain why quantitative researchers typically use deductive logic and qualitative researchers typically use inductive logic.

8. List and describe the steps that researchers should follow when developing a survey or questionnaire.

9. Compare and contrast the advantages and limitations associated with using mail surveys, personal interviews, and telephone interviews.

10. Identify and discuss the two primary characteristics of a good test or inventory.

11. List and describe the methods used by researchers to accumulate evidence to make decisions regarding the validity of a test or an inventory.

12. List and describe the methods used by researchers to assess the reliability of a test or an inventory.

Expanded Outline

I.   Descriptive Methods
          Archival and Previously Recorded Sources of Data

               Potential Problems
               Comparisons with the Experimental Method
          Observational Techniques
               Case Studies
               Naturalistic Observation

                    Reactance or Reactivity Effect
                    Hawthorne Effect
               Participant Observation
                    Ethnography

               Clinical Perspective
               Choosing Behaviors and Recording Techniques
                    Time Sampling
                    Situation Sampling
          Using More than One Observer: Interobserver Reliability

II.   Qualitative Research

III.   Correlational Studies

IV.   Review Summary

V.   Check Your Progress

VI.   Ex Post Facto Studies

VII.   Surveys, Questionnaires, Tests, and Inventories
          Surveys and Questionnaires
               Types of Surveys
                    Descriptive Survey
                    Analytic Survey
                    Pilot Testing

Developing a Good Survey or Questionnaire
    Yes-No Questions
    Forced Alternative Questions
    Multiple-Choice Questions
    Likert-type Scales
    Open Ended Questions

    Demographic Data
Mail Surveys

    Personal Interviews
    Telephone Interviews
Tests and Inventories
    Characteristics of Good Tests and Inventories
    Validity
        Content Validity
        Interrater Reliability
        Concurrent Validity
        Criterion Validity
    Reliability
        Test-Retest Procedure

        Split-Half Technique
    Types of Tests and Inventories
        Achievement Test
        Aptitude Test
        Personality Test or Inventory

VIII.   Review Summary

IX.   Check Your Progress

X.   Key Terms

XI.   Looking Ahead

Practice Exam

Multiple Choice

Identify the letter of the choice that best completes the statement or answers the question.

_____ 1. Psychologists who answer their research questions by using previously recorded data are using

    a. a manipulated IV.   c. a recorded DV.
    b. archival data.   d. none of the above

_____ 2. Which of the following are potential problems associated with using previously recorded data?

    a. you may not know the source of the data   d. a and c
    b. the participants were selective in what they wrote   e. all of the above
    c. the survival of the data records

_____ 3. Paul Broca's research method for studying his patient named "Tan" was more closely related to _____ research.

    a. survey   c. naturalistic observation
    b. case study   d. archival

_____ 4. The goals associated with using naturalistic observation methods include

    a. describing behavior as it naturally occurs.   d. a and b
    b. the researcher becoming a part of the study.   e. a and c
    c. describing the variables and their relationships.

_____ 5. Which of the following is <u>not</u> a common observational technique?

    a. case studies   c. archival studies
    b. naturalistic observation   d. surveys

_____ 6. The _____ refers to researchers who primarily observe a situation, but also interact with others; whereas the _____ refers to researchers who become a part of the culture by working and interacting extensively with the others.

    a. observer as participant,   b. participant as observer,
        participant as observer   observer as participant

___ 7. The advantages of situation sampling include

 a. determining whether the behavior in question changes   d. a and c
    as a function of the context in which it was observed.
 b. increasing the researchers ability to generalize.   e. all of the above
 c. observing different participants in different situations.

___ 8. All of the following are considered to be limitations of participant observation
       except:

 a. the observer may lose objectivity   c. the ability to make cause-and-effect
                                                   statements
 b. the observer becomes close   d. a long period of time may be necessary
    to the participant group(s)   for the observer to be accepted

___ 9. When two or more individuals observe the same behavior, researchers can
       compare their observational records for agreement or disagreement by examining

 a. content validity.   c. demographic data.
 b. criterion validity.   d. interrater reliability.

___ 10. Qualitative researchers prefer to use _____ logic; whereas quantitative
        researchers typically prefer to use _____ logic.

 a. inductive, reductive   c. deductive, inductive
 b. inductive, deductive   d. deductive, reductive

___ 11. Which represents a true distinction between the clinical perspective (CP) and the
        participant observation (PO) methods.

 a. A clinician's goal is understanding,   c. a participant observer can remain
    whereas a participant observer's goal   passive and a clinician must intervene
    is helping
 b. unlike clinicians, participant observers   d. a clinician deals with a broad picture
    cannot be unobtrusive because they   of behavior, while the participant
    participate in the situation   observer deals with a narrow picture of
                                                         behavior

___ 12. When researchers are working with IVs that they cannot or do not manipulate,
        they are conducting a(n)

 a. one-way ANOVA.   c. ex post facto study.
 b. t-test.   d. z-test.

____ 13. _____ refers to the evaluation of an analytic survey performed in advance of the complete research project.

    a. Pre-study testing   c. A post hoc study
    b. Pilot testing   d. An ex post facto study

____ 14. A survey or questionnaire in which individuals are asked to respond to a stimulus by selecting a response choice (usually 3 to 7 choices) are called

    a. Likert-type scales.   c. yes-no scales.
    b. forced alternative scales.   d. open ended scales.

____ 15. During data collection researchers obtain relevant information about their participants such as age, gender, income, and major. What is this information typically called?

    a. demographic data   c. descriptive data
    b. inferential data   d. personal data

____ 16. The final step in developing a survey or questionnaire requires determining

    a. the administration procedures.   c. the items for your survey.
    b. how the data will be obtained.   d. when to pilot test your survey.

____ 17. A typical participant response rate for mail surveys is

    a. 5-10 percent.   c. 20-25 percent.
    b. 10-15 percent.   d. 25-30 percent.

____ 18. When a trained interviewer administers a survey in a respondent's home it is not uncommon to have a _____ percent completion rate.

    a. 20   c. 60
    b. 30   d. 90

____ 19. Which of the following is not considered to be a general type of test or inventory?

    a. achievement   c. aptitude
    b. opinion poll   d. personality

____ 20. A good test or inventory should be

    a. reliable, but not valid.   c. reliable and valid.
    b. valid, but not reliable.   d. neither reliable nor valid.

_____ 21. When a test score will be compared with a future outcome, the researcher is attempting to examine _____ validity.

    a. content   c. concurrent
    b. criterion   d. face

_____ 22. Reliability is typically assessed by using

    a. test-retest procedures.   d. both a and b
    b. split-half procedures.   e. both b and c
    c. post hoc procedures.

_____ 23. Problems associated with using test-retest reliability procedures include

    a. history effects.   d. only b and c
    b. test familiarity.   e. all of the above
    c. lengthy time periods.

_____ 24. _____ tests are administered when an evaluation of an individual's level of mastery or competence is desired; whereas _____ tests are administered to assess an individual's ability or skill in a particular situation.

    a. Achievement, aptitude   c. Personality, achievement
    b. Aptitude, personality   d. Aptitude, achievement

_____ 25. You are planning to take the Graduate Records Exam (GRE) as part of an application requirement for graduate school. What type of test is the GRE?

    a. achievement   c. personality
    b. aptitude   d. mastery or competence

Matching

a. observer as participant

b. Hawthorne effect

c. content validity

d. participant as observer

e. demographic data

f. Graduate Records Exam

g. interrater reliability

h. interobserver reliability

i. concurrent validity

j. criterion validity

k. the bar exam for lawyers

_____ 1. a type of achievement test

_____ 2. a researcher who primarily observers a situation, but who interacts with others

_____ 3. compares scores on a test with another measure of the designated trait

_____ 4. synonymous with reactance or the reactivity effect

_____ 5. a type of aptitude test

_____ 6. measures the agreement of judges by examining the content validity of tests

_____ 7. examines the extent that test items represent the content they were intended to represent

_____ 8. measures the extent to which observers agree

_____ 9. a researcher who becomes part of a culture by working and interacting extensively with others

_____ 10. participant information such as gender, age, and education level

_____ 11. compares a score on a test with a future score on another test

True/False

Indicate whether the sentence or statement is true or false.

_____ 1. Researchers can only speculate about cause-and-effect relationships when using descriptive research methods.

_____ 2. Observational techniques involve the direct manipulation of a variable.

_____ 3. Researchers who intensively observe and record the behavior of participants over time are using archival data collection procedures.

_____ 4. The "observer as participant" refers to a researcher who primarily observes a situation, but who interacts with others.

_____ 5. Qualitative researchers typically rely on deductive logic.

_____ 6. Qualitative researchers typically analyze their data simultaneously with data collection, data interpretation, and narrative report writing.

_____ 7. Analytic surveys seek to determine relevant variables and their relationships.

_____ 8. A test is only considered to be reliable when it measures what it was intended to measure.

_____ 9. Validity only refers to tests or inventories that consistently measure the same individuals over repeated administrations.

_____ 10. Aptitude tests measure an individual's ability or skill in a particular setting.

Short Answer

1. Discuss the purpose for using descriptive research methods. Explain the potential problems associated with using archival or previously recorded data sources.

2. Compare and contrast the most common observational techniques used by researchers.

3. Explain why researchers should be unobtrusive when using naturalistic observation research methods.

4. List and describe the limitations associated with using participant observations.

5. Discuss the differences between the clinical perspective and participant observation research methods.

6. Summarize the purpose for using interrater reliability measures.

7. Explain why quantitative researchers typically use deductive logic and qualitative researchers typically use inductive logic.

8. List the steps that researchers should follow when developing a survey or questionnaire.

9. Compare and contrast the advantages and limitations associated with using mail surveys, personal interviews, and telephone interviews.

10. Discuss the two primary characteristics of a good test or inventory.

11. List and describe the methods used by researchers to accumulate evidence to make decisions regarding the validity of a test or an inventory.

12. List and describe the methods used by researchers to assess the reliability of a test or an inventory. What are the benefits and limitations associated with using these methods?

Answers for the multiple choice, matching, and true/false items

Multiple Choice		Matching		True and False	
1.	B	1.	K	1.	T
2.	D	2.	A	2.	F
3.	B	3.	I	3.	F
4.	E	4.	B	4.	T
5.	C	5.	F	5.	F
6.	A	6.	G	6.	T
7.	E	7.	C	7.	T
8.	B	8.	H	8.	F
9.	D	9.	D	9.	F
10.	B	10.	E	10.	T
11.	C	11.	J		
12.	C				
13.	B				
14.	A				
15.	A				
16.	A				
17.	D				
18.	D				
19.	B				
20.	C				
21.	B				
22.	D				
23.	D				
24.	A				
25.	B				

Chapter 20

Writing, Assembling, and Publishing an APA-Format Research Report

Learning Objectives

By the end of this chapter you should be able to:

1. Sequentially list the sections of an APA formatted paper.

2. Describe the contents that should appear on the title page.

3. Explain the purpose of an abstract.

4. Explain why the authors use the analogy of a "funnel" to describe a good introduction section.

5. List and describe the typical components of a method section.

6. Discuss the purpose of a results section and explain when it is appropriate to use tables and figures to communicate your results.

7. Summarize the information that should be presented in a discussion section.

8. Describe the components of and organizational rules for constructing a reference section.

9. Explain the purpose and typical content of an appendix.

10. List and describe the three strategies, as recommended by the APA Publication Manual, for improving writing styles.

11. Provide an example of the appropriate use of the words that, which, since, and while in the context of research articles.

12. Distinguish between the terms "active" and "passive" voice and explain why researchers should strive to use the "active voice" when writing a research report.

13. Explain why researchers should use unbiased language when writing research articles.

Expanded Outline

I.   What is APA Format?

II.   Sections of the APA Format Paper
           Title Page
                Manuscript Page Header
                Running Head
           Abstract

           Introduction

                Thesis Statement
                Citation
                Reference
                Reference Section

                Unbiased Language
           Method
                Participant Subsection
                      Level 3 Heading
                Apparatus, Materials, or Testing Instruments
                      Materials Subsection
                      Apparatus Subsection
                      Testing instrument(s) Subsection
                Procedure Subsection

           Results
                Inferential Statistics

                Descriptive Statistics
                Complementary Information
                      Figures
                      Tables

           Discussion
                Restating Results
                Comparing Results to Previous Research
                Interpreting the Results

Practice Exam

Multiple Choice

Identify the letter of the choice that best completes the statement or answers the question.

_____ 1.   Psychological researchers depend on the _____ guidelines to prepare and communicate the results of their research.

    a. American Educational   c. APA Publication Manual
        Research Association
    b. APA Ethical Research   d. APA Thesaurus

_____ 2.   Authors use _____ to organize and divide the APA formatted paper into distinct sections.

    a. titles   c. indentations
    b. headings   d. bolding

_____ 3.   The most current edition of the APA Publication Manual is the

    a. 2nd edition.   c. 4th edition.
    b. 3rd edition.   d. 5th edition.

_____ 4.   Which of the following is <u>not</u> found on the title page?

    a. abstract   c. running head
    b. author affiliation(s)   d. page header

_____ 5. The manuscript page header consists of the

    a. first word of the title.   c. first two or three words of the title.
    b. longest word of the title.   d. the complete title.

_____ 6. The running head should be a maximum of _____ characters, including letters punctuation, and spacing between words.

    a. 20   c. 40
    b. 30   d. 50

_____ 7. When a research article has more than one author, the authors' names appear in

    a. alphabetical order.   c. order of educational experience.
    b. order of the importance of their contributions.   d. no certain order.

_____ 8. According to the APA Publication Manual, the title of a manuscript should be _____ words in length.

    a. 4-6   c. 10-12
    b. 7-9   d. there is no page length requirement

_____ 9. An abstract of a research article

    a. briefly describes your paper.   c. introduces the authors.
    b. organizes your tables and figures.   d. organizes the references.

_____ 10. Which citation does not follow APA format?

    a. Davis, Smith, and Juve (2005)   c. (Davis, Smith, & Juve, 2005)
    b. (Davis, Smith, and Juve, 2005)   d. all citations are accurate

_____ 11. The method section is typically comprised of

    a. participants.   c. procedures.
    b. apparatus.   d. all of the above

_____ 12. Martillë is writing the introduction of her research paper and finds a citation with four authors. If she is citing this reference in the introduction of her paper for the first time she should include:

    a. the last name of the primary author followed by et al.   c. all four authors
    b. only the last name of the primary author   d. the first three authors

_____ 13. When researchers present the inferential statistical results of their research they <u>must</u> include the

   a. degrees of freedom.   d. effect size.
   b. test statistic.   e. all of the above
   c. probability.

_____ 14. Figures are more appropriate than tables for presenting

   a. standard deviations.   c. means.
   b. interactions.   d. all of the above

_____ 15. Which subsection is typically the longest component of the method section?

   a. procedure   c. participants
   b. instrumentation   d. apparatus

_____ 16. When citing a reference from a periodical article the information appearing immediately after the author(s) of the article reflects the

   a. title of the article.   c. the volume number.
   b. title of the periodical.   d. date of the article.

_____ 17. The following reference format typically reflects a citation from a

   Author, A. A. (date). _Title of the work_. Location: Publisher.

   a. periodical article.   c. chapter from an edited book.
   b. book.   d. world wide web source.

_____ 18. Tiger is writing the results section of his research paper. He included the following information:

   "The data analysis revealed a significant main effect of race such that African-American golfers ($M = 4.32$, $SD = .91$) were rated more positively than were Caucasian golfers ($M = 3.76$, $SD = 1.00$), $F(1, 92) = 10.42$, $p = .002$, and $\eta^2 = .10$."

   What information in this example is considered to be a descriptive statistic?

   a. M and SD d. p and η^2
   b. SD and F e. both a and b
   c. F and p

___ 19. **Using the data from question 18**, what information is considered to be an inferential statistic?

    a. M and SD c. F and p
    b. SD and F d. all of the above

___ 20. When citing _____ sources it is very important to include the day and year that you retrieved the information, given that the cited information can be frequently updated.

    a. periodical article   c. world wide web
    b. book   d. edited book chapter

___ 21. Clauses beginning with _____ are termed restrictive clauses and should be essential to the meaning of the sentence; whereas clauses beginning with _____ can be either restrictive or nonrestrictive (i.e., simply adding additional information).

    a. that, which   c. that, since
    b. which, that   d. while, which

___ 22. Which of the following uses of the words "since" and/or "while" is acceptable according to the APA Publication Manual.

    a. many different cars have evolved <u>since</u>   c. <u>while</u> some researchers believe in
        Ford's original version of the model T   Freud's theories, others have
                                                     rejected his theories.

    b. <u>since</u> the pro golfers performed better   d. both a and b
        than the amateur golfers, the researcher
        concluded that they were better athletes.

Matching

a. level 3 heading

b. method section

c. plagiarism

d. abstract

e. level 1 heading

f. manuscript page header

g. level 4 heading

h. figure

i. introduction

j. table

k. running head

l. headings

_____ 1.   a pictorial representation of the results

_____ 2.   using other research or ideas without providing credit to the original source

_____ 3.   consists of the first two to three words of the title

_____ 4.   a section title that is left justified, italicized, has the first letter of each major word capitalized, and occupies a line by itself

_____ 5.   a shortened title printed at the top of alternative pages in published journals

_____ 6.   a centered section title in which the first letters of major words are capitalized and occupies a line by itself

_____ 7.   section of a paper containing the experimental hypothesis

_____ 8.   titles helping readers understand the organization and content of research papers

_____ 9.   section of a paper containing information about the participants and materials

_____ 10.   a chart containing an array of descriptive statistics

_____ 11.   brief description of a research paper

_____ 12.   a section title that is indented five spaces, italicized, has only the first word capitalized, and ends with a period

True/False

Indicate whether the sentence or statement is true or false.

_____ 1. Headings are used in the reference section to separate authors.

_____ 2. The authors note represents the first page of an APA formatted paper.

_____ 3. The manuscript page header appears on all pages of the paper <u>except</u> for the figures.

_____ 4. The running head should not exceed 30 characters including letters, punctuation, and spacing between words.

_____ 5. The introduction section of a research paper begins on the third page.

_____ 6. The method section is typically comprised of three subsections: participants, apparatus, and procedures.

_____ 7. The participant subsection is typically the longest of the three components of the method section.

_____ 8. The advantage of using figures instead of tables is that standard deviations can be included in figures.

_____ 9. The reference section includes a list of every source that researchers read when they were writing their introduction and planning their experiment.

_____ 10. If you have more than one article by the same first-author, you should alphabetize the citations by using the name of the second author.

Short Answer

1. Sequentially list the sections of an APA formatted paper. Do any of these sections appear on a separate page?

2. List the content that should appear on the title page.

3. Explain the purpose of an abstract. What is the suggested length for an abstract?

4. Summarize why the authors use the analogy of a "funnel" to describe a good introduction section.

5. List and describe the typical components of a method section.

6. Discuss the purpose of a results section. Explain when it is appropriate to use tables and figures to communicate your results.

7. Summarize the information that should be presented in a discussion section. According to the 4th edition of the APA Publication Manual (used instead of the 5th edition because the authors prefer the simpler and clearer guidelines), what are the three questions that should guide the organization and writing of a discussion section?

8. Summarize the components of and organizational rules for constructing a reference section. Describe the similarities and differences associated with citing periodical articles, books, chapters from edited books, world wide web sources, and other references. Provide a brief example of each type of reference.

9. Explain the purpose of an appendix. What content would you typically find in an appendix?

10. List and describe the three strategies, as recommended by the APA Publication manual, for improving writing styles.

11. Provide an example of an appropriate use of the following words in the context of research articles: that, which, since, and while.

12. Distinguish between the terms "active" and "passive" voice. Explain why researchers should strive to use the "active voice" when writing a research report.

13. Explain why researchers should use unbiased language in their writing. Provide an example sentence that demonstrates the use of biased and unbiased language.

Answers for the multiple choice, matching, and true/false items

Multiple Choice		Matching		True and False	
1.	C	1.	H	1.	F
2.	B	2.	C	2.	F
3.	D	3.	F	3.	T
4.	A	4.	A	4.	F
5.	C	5.	K	5.	T
6.	D	6.	E	6.	T
7.	B	7.	I	7.	F
8.	C	8.	L	8.	F
9.	A	9.	B	9.	F
10.	B	10.	J	10.	T
11.	D	11.	D		
12.	C	12.	G		
13.	E				
14.	B				
15.	A				
16.	D				
17.	B				
18.	A				
19.	C				
20.	C				
21.	A				
22.	A